Gold Metallurgy and the Environment

Gold Metallurgy and the Environment

By
Sadia Ilyas and Jae-chun Lee

CRC Press
Taylor & Francis Group
Boca Raton London New York

CRC Press is an imprint of the
Taylor & Francis Group, an **informa** business

CRC Press
Taylor & Francis Group
6000 Broken Sound Parkway NW, Suite 300
Boca Raton, FL 33487-2742

First issued in paperback 2020

ISBN-13: 978-0-367-57207-5 (pbk)
ISBN-13: 978-1-138-55685-0 (hbk)

Library of Congress Cataloging-in-Publication Data

Names: Ilyas, Sadia, editor. | Lee, Jae-Chun, editor.
Title: Gold metallurgy and the environment/[contributions by] Sadia Ilyas, Jae-Chun Lee.
Description: Boca Raton: Taylor & Francis, CRC Press, 2018. | Includes bibliographical references and index.
Identifiers: LCCN 2017049455 | ISBN 9781138556850 (hardback : alk. paper) | ISBN 9781315150475 (ebook)
Subjects: LCSH: Gold--Metallurgy. | Extraction (Chemistry)--Environmental aspects.
Classification: LCC TN760 .G685 2018 | DDC 669/.9622--dc23
LC record available at https://lccn.loc.gov/2017049455

Visit the Taylor & Francis Web site at
http://www.taylorandfrancis.com

and the CRC Press Web site at
http://www.crcpress.com

This book is dedicated to my beloved parents whose footprints of grace

have kept me going further in life and continuously enlighten me.

Sadia Ilyas

Contents

Acknowledgement

It is a great privilege to express personal thanks to Dr. Rajiv Ranjan Srivastava, whose enthusiasm has always provided positive encouragement. His generous, humane, and endless support have helped immensely to shape this book in its form.

Preface

This book is written with the intention of providing comprehensive coverage on the metallurgical exploitations of gold from its divergent ores, discussing their effects on the environment, and exploring the possibilities for solving those problems. Some good books on the topics related to gold metallurgy are available; however, this book also covers environmental issues. The aspects of artisanal gold mining/gold amalgamation are therefore given proper coverage; after which, in the last chapter, human perceptions towards gold metallurgy including some case studies, are provided, along with a fair discussion on the fundamentals involved in processing the gold-bearing ores.

The depletion of high-grade ores is a serious issue, which necessitates the exploitation of gold-bearing refractory ores creating more obstacles in processing them with the use of energy-intensive/chemicals. Hydrometallurgical techniques are therefore of greater importance, and we attempted to relate the aqueous chemistry involved in the process fundamentals. The use of microbial activities, specifically for refractory ores, is therefore given attention from an environmental and cost-benefit point of view. Although all of the extraction techniques are described, because cyanidation is the most prominent, it is discussed in greater detail, as well as the toxicity that accompanies it. Other than the chemical alternatives, microbial alternatives including bio-cyanidation techniques are discussed in a separate chapter. So that it is easy to understand, the full processing of each route and the recovery is discussed in the same chapter with the particular source (medium) of the processing leach liquor. Considering the environmental issues, treatment of cyanide effluents by the meaning of both chemical and microbial routes is given equal importance. We have kept the book focused on metallurgical exploitation and environmental impacts, hence only a limited part of mining and mineralogy is considered. Throughout the book, the practical applications of chemical principles and techniques are emphasized, with their industrial applications as well.

We believe that the book will be beneficial to professionals involved in the exploitation of this precious metal, and of particular interest to students and the research community.

Sadia Ilyas *(Faisalabad, Pakistan)*
Jae-chun Lee *(Daejeon, South Korea)*
January 23, 2018

Editors

Sadia Ilyas is currently working as an assistant professor in the Department of Chemistry, at the University of Agriculture Faisalabad in Pakistan. After receiving her M.Phil. and Ph.D. in Inorganic Chemistry from Bahauddin Zakariya University (Multan) and University of Agriculture (Faisalabad), respectively; she served in China as senior researcher at the School of Chemical Engineering and Pharmacy (Wuhan) and in Korea as a postdoctoral fellow at the Korea Institute of Geoscience and Mineral Resources, and in Pakistan as assistant professor at GC University Faisalabad. Besides the academic lecture in inorganic chemistry and hydrometallurgy, her research deals with hydrometallurgical exploitation of metals from primary and secondary sources using microbial and chemical activities. She has published more than 30 research articles in peer-reviewed international journals and is also an author in three books, nine contributed book chapters, textbooks, and laboratory manuals.

Jae-chun Lee is currently a distinguished principal researcher in the Mineral Resources Research Division at the Korea Institute of Geoscience and Mineral Resources (KIGAM) and an adjunct professor in the Department of Resources Recycling at the Korea University of Science & Technology. Dr. Lee received his B.S. in metallurgical engineering and M.S. and Ph.D. in Hydrometallurgy from Hanyang University, Seoul, South Korea. His research deals with leaching, separation, and purification of metals from primary and secondary resources. His current research focuses on the recycling of valuable metals from urban mines by hydrometallurgical routes. Dr. Lee has published around 300 research articles in several international peer-reviewed journals and contributed to three book chapters.

1

Gold Ore Processing and Environmental Impacts: An Introduction

Sadia Ilyas*

1.1 Overview

Gold is a precious metal; its worldwide influences proven in our economy, technology, and traditional values. Originated from the Latin word aurum, Au is the symbolic representation for this precious metal with the atomic number 79. It is a ruddy yellow, delicate, bendable, splendid, and thick metal in the pure form. The unique ruddy yellow appearance is due to the characteristic structure of gold, which readily absorbs electromagnetic radiation below 560 nm but reflects radiation above 560 nm. It is one of the minimum receptive substance components and is strong under standard conditions. Along with the other coinage metals, copper and silver, gold is a member of group IB of the periodic table, which is soluble under oxidative environment in aqua regia, cyanide, thiosulfate, thiourea, and halide solutions. It has face-centred cubic (fcc) crystalline structure with electronic configuration $[Xe]4f^{14}5d^{10}6s^1$. The various characteristics of this precious metal are summarized in Table 1.1. Its chemical stability and high resistance to oxidation allow it to be the prominent noble metal. Notably, the electrical resistivity of gold (22.1 nΩ/m) is lesser than copper (17 nΩ/m) and silver (16 nΩ/m), but the latter two metals corrode at atmospheric conditions. Gold is present in the form of grains, chunks in alluvial stories, rocks, and veins in its free state. Gold is often associated with silver as electrum and with minerals like (chalco/arseno)-pyrites. Further, gold is also blended with tellurium as gold tellurides in the form of minerals and anode slimes of copper refineries.

* Mineral and Material Chemistry Lab, Department of Chemistry, University of Agriculture Faisalabad, Pakistan.

TABLE 1.1

Typical Characteristics of the Precious Metal, Gold

Atomic number	79
Electronic configuration	$[Xe]4f^{14}5d^{10}6s^1$
Mass number of natural isotope	197
Atomic weight	197.2
Density at 20 °C, g/cc	19.3
Atomic volume of solid, cc	10.22
Melting point, °C	1063
Boiling point, °C	2600
Ionization potential, eV	9.223
Atomic radii, pm	174

1.2 A Historical Perspective

From ancient time to the present era of the techno-world, gold has been esteemed by humans. Egypt is the oldest gold-delivering country as evident from the map of a gold mine, supposedly drawn in 1320–1200 BC. Coptos in Wadi Hammamat was the world's first gold-blast town where the gold bearing veins and rocks washed out the precious metal. Under Egyptian administration, approximately hundreds of underground mines were explored in Nubia during 1300 BC. In the Middle East, Egypt turned into an overwhelming force with the best gold treasury in their antiquated earth. Via the Red Sea, old Egyptians exchanged their circumstances because they did not have their own critical port at the Mediterranean. In the south-eastern side of the Black Sea, the date of gold exploitation is said to coincide with the time of Midas that was vital to establish the world's earliest coinage in Lydia (at ~610 BC). From the 6th century BC, the Chu (state) circulated a square gold coin, Ying Yuan. A large scale operation of gold in Rome was supposed to have started from 25 BC onwards (in Hispania) and from 106 AD onwards (in Dacia). Some of the smaller deposits in Britain (like placer and hard-rock deposits) at Dolaucothi were also exploited by them.

1.2.1 Mining

In 2016, 3,236 t of gold was mined, with a 5% increase in supply at present. This is far ahead of what Herodotus (484–425 BC) alluded to as a few extraordinary gold-mining focuses in Asia Minor, and Strabo's (63 BC) notice of gold-mining in a wide range of spots. Pliny (23–79 AD) gives many points of interest of antiquated in-situ mining, which was broad. The Romans had a small amount of the metal in their locales; however, their armed forces' endeavours brought them significant sums as goods. They additionally abused the mineral abundance of nations they had conquered, particularly Spain, where up to 40,000 workers were utilized in the mining area. The state's amassing of gold was huge; however, amid brute attacks and the fall of the realm this gold was spread and gold mining morphed in the Middle Ages. In spite of the fact that the conquistadors found an exceptional mining industry in Central America, their endeavours to build a gold generation were to a great extent unsuccessful on the grounds that the greater part of the discoveries consisted of silver. Around 1750, gold was exploited on a noteworthy extent on the eastern slants of the Ural Mountains. During the mid-19th century, alluvial gold was found in Siberia and Coloma, California. Coloma is around 50km southeast of Sacramento on the slants of the Sierra Nevada. A chronicled presentation gold stores likewise originated in Western Canada (1896), New Zealand and Western Australia (1892), Colorado (1875), Nevada (1859), and Eastern Australia (1851). In any case, those stores rapidly lost quite a bit of their significance. In 1885 their most grounded stimulus was given to gold creation via disclosure of the gold fields of the Witwatersrand in South Africa. South African gold soon became a prime location on the planet showcase. Creation developed consistently aside from a tiny intrusion of the Boer War (1899–1902). In the Middle Ages gold mining in the Gold Coast(Ghana) started to assume an unobtrusive part in the twentieth century, despite the fact that the stores were identified. Near the Thompson River, around the Fraser River, and then on Williams and Lightning Streams, the alluvial gold sources were found between 1858 and 1885. In 1911, the region of British Columbia was the major gold exploited area in the Canadian region and domains for a long time, yet with the disclosure of the Kirkland Lake stores, and in 1912 the opening up of the Porcupine locale, Ontario was in the front of the rest of the competition from that point onward.

1.2.2 Gold and Alchemy

The medieval chemists identified gold as a flawless metal. Chemists distinguished it by the sun with the righteousness of its splendid yellow shading, and this was given the image of a hover with a speck inside. Gold was so valuable that in every time man did everything they could to find it in nature. It is not astounding, that people have tried to change over different metals into gold. The philosopher's stones were known as specialists that converted base metal to precious gold. Notwithstanding its convertory control, stones were accepted to have the properties of an all-inclusive drug for life span and everlasting status. Catalytic standards additionally have discovered their way into present day mental thoughts, notably by the Swiss specialist Carl Jung (1875–1961). The origin of speculative chemistry was known as Antiquated Egypt I. Zosimos (350–420), who was educated in Alexandria, is the earliest essayist identified as having honed speculative chemistry. In view of the absence of information of the organization of regular substances, a few chemists saw numerous conventional synthetic reactions as transmutor. For instance, testimony of Cu on press metal set in blue vitriol ($CuSO_4$), a response well-known in Roman circumstances, where it was expected by a few to be a conversion of Fe into Cu in anticipation of the late recovery. During that time Au-production had been both supported and restricted by rulers and the Church. Amid medieval circumstances European lords and sovereigns bolstered chemists at their courts wanting to gain riches through their work.

1.3 Occurrence of Gold

With a crustal abundance of 3–5 parts per billion (ppb), this precious metal in the earth is supposed to come via two sources:

i. The supernova event: the collision of two or more neutron stars in space created several nuclear synthesis processes, which leads to the formation of a wide range of heavy elements. During the condensation of heavy elements and forming solar system, the elements which fell to the molten earth could end up sinking to the core (Seeger et al., 1965).

ii. The asteroid bombardment: the impact created around 4 billion years ago could enrich the earth's crust and upper mantle through the bombardment of gold from the gold-bearing asteroids (Willbold et al., 2011). The widespread distribution of gold throughout the earth's crust has been also explained by the asteroid bombardment model.

Besides the previously mentioned source of gold in earth's crust, the formation of gold-bearing ores can be *Endogenetic* by hydrothermal process and *Exogenetic* by weathering effects of wind and/or water erosion of gold-bearing hard rock deposits. The heating of highly mineralized ground water caused by igneous intrusions enables water to adsorb and/or dissolve metals from surrounding rock formations; flow via rock fractures can produce veins which deposit microscopic particles of native gold embedded within calcite, quartz, and other minerals (Hamburger et al., 2010). A typical hydrothermal gold deposition (*Endogenetic* and *Exogenetic*) system is shown in Figure 1.1. Whereas, the erosion which drove *Exogenetic* eventually causes the alluvial, or placer deposits, located in the river system and coastal regions (Kettell, 1982). In a stream when the water slowly flows down, the higher density gold ($19.32\,g/cm^3$) drops out first, leaving the lighter rock minerals ($\sim2.7\,g/cm^3$) to continue to flow to concentrate gold. The placer deposits were the first to be exploited from the rivers, using the panning and sluicing techniques followed by the smelting.

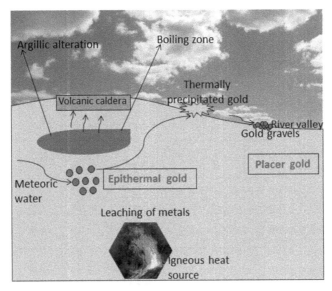

FIGURE 1.1
A typical representation of the hydrothermal deposition (*Endogenetic* and *Exogenetic*) of gold.

1.4 Gold Mineralogy

Most of the gold occurs as local metal, and most of the minerals are naturally connected to it. Silver often coexists with gold; when it exists with >20% silver it is named as the "electrum." Other gold amalgams are uncommon and for the most part limited to particular metals; for instance, the two gold copper composites: Cu_3Au (auricupride) and $[AuCu]_4$ (tetra-auricupride) are considered high gold-review porphyry copper minerals (Kemess, BC, Canada). Some important ones are FeS_2 (pyrite), PbS (galena), ZnS (zincblende), FeAsS (Arsenopyrite), Sb_2S_3 (stibnite), FeS (pyrrhotite), and $CuFeS_2$ (chalcopyrite). Different Se minerals and Fe_3O_4 (magnetite) may likewise be available. In Witwatersrand, South Africa, UO_2 uraninite, and to a lesser degree, thucholite (a variable blend of hydrocarbons, uraninite, and sulphides) are related to the gold bearing mineral. Uranium was recuperated as a result of gold exploitation. Carbonaceous matter always relates to certain gold minerals. Gold has partiality for Te with two known minerals, the calaverite, $AuTe_2$ and sylvanite $(Ag,Au)Te_2$. It occurs with Pd as porpezite, and with Rd as resin-rhodite. In placer stores, it might be available as small grains or huge pieces. In specific metals knowing headstrong minerals, gold is related with sulfidic minerals to a great degree of a finely separated state. Petrovskaya (1987) worked on revealing the tendency of elements to associate with gold in minerals and ore bodies; this information is presented in Figure 1.2. Notably, the lode (vein) and placer deposits are mainly recognized auriferous deposits;

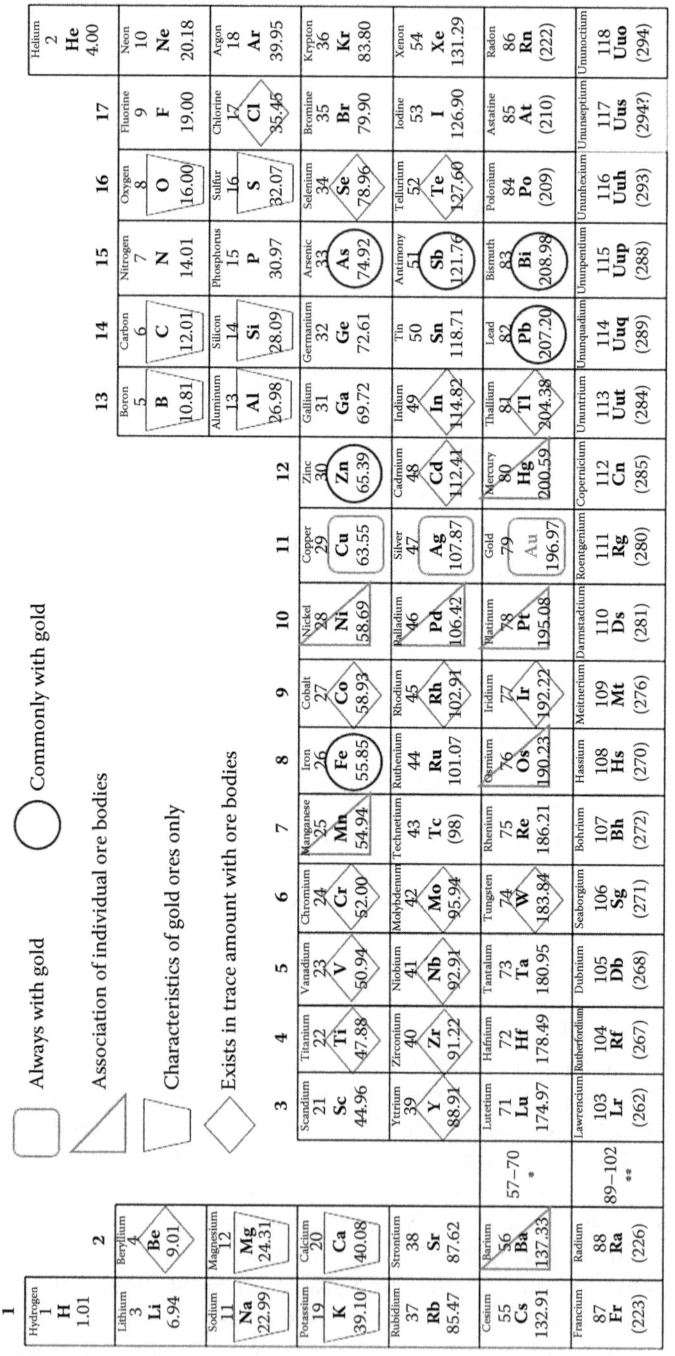

FIGURE 1.2

The geochemical association of elements with gold in their minerals and ore-bodies.

however, the quartz-pebble deposits supplementing ~50% of worldwide gold production are classified as modified paleo-placer (Yannopoulos, 1991).

1.4.1 Mineral Processing of Gold-Bearing Ores

1.4.1.1 Gravity Concentration

Gravity focus techniques are sound in light of the thickness distinction between the constituent minerals of a metal for their partition (Burt and Mills, 1984). The movement of two mineral particles of various densities in a liquid will be extraordinary. The heavier mineral will quickly settle down, while the light one will be suspended in the liquid or washed away by the moving liquid. On account of the colossal thickness contrast between gold (thickness 19.3) and the gangue minerals (normal thickness 2.7–3.5) which are available in the metal, gravity division operations are generally utilized for gold focus. The mineral particles must be adequately freed for a decent partition. By and large, freedom of the constituent minerals is accomplished by comminution and crushing. Over pounding would be negative to both the partition productivity and vitality utilization. Gravity division effectiveness altogether diminishes with diminishing particles measure, however new innovative upgrades of gravity fixation gadgets have prompted enhanced partition productivity while treating fine-grained minerals. Practically speaking, the coarser-grained particles go from a couple of mm (e.g., 2mm) down to roughly 40m. Such great freed gold particles are reasonable for gravity focus, which is then favoured for a quicker recuperation of qualities. Gold from placer stores is recouped for the most part by utilizing gravity focus strategies. The most utilized gravity fixation gadgets today incorporate conduits, shaking tables, and dances. They all endeavour the liquid (water) development, streaming by floodgates and tables or throbbing by dances, to upgrade the partition of various minerals. Additionally, created mechanical gear like the "Knelson concentrator" even makes utilization of outward drive. The divergent separators can work productively at little molecule measure down to 30m (Richards, 1988). Gravity focus stream sheets for the most part incorporate various blends into the accompanying operations: comminution, wet crushing, granulometric characterization of the mineral particles, multi-arrange fixation, think recuperation, and deposit dispose of. Infrequently, the gravity gold concentrates are tidied up by amalgamation (Shoemaker, 1984). Amid this procedure, the free gold particles are wetted by mercury and shaped into an amalgam-sort Richard compound at room temperature and would thus be able to be isolated from the greater part of contaminations. From that point, mercury is isolated from gold by refining. It is continuously reused subsequent to cooling. The strategy was once broadly utilized as a part of gold metals preparation; however, it is currently practically deserted in light of the danger to wellbeing that results from dealing with mercury.

Recuperation techniques for GRG and gold transporters change, as no one but GRG can be significantly recouped at less weight recuperation into the focus, or yield 0.1%, run of the mill of the sort of semi-consistent units utilized today, while gold bearers – for example, FeS and FeAsS – are recuperated by constant units able to do considerably higher yields, which ordinarily coordinate or somewhat surpass the sulphide substance of the treated stream. Gravity detachment has been used in gold plants as the essential recuperation system, or then again in front of other downstream procedures. For example, buoyancy and cyanidation have been used since the commencement of mineral preparing. In 1966 at Australian gold mines a substantial dependence on gravity in the crushing is ruled by the dance, strake, and shaking (Elvey and Woodcock, 1966). The recuperation of free and sulphide (FeS, FeAsS and Te) related Au from the essential pounding circuit is highlighted in every one of these establishments. In Australia crushing circuit is essential for all intents and purposes in fused gravity recuperation. Gold is recouped from concentrate by use of gravity table via cyanidation and amalgamation and recovery treatment went on. The utilization of amalgamation was highlighted in the mining area; however, it has been eliminated because of wellbeing and natural issues. As of not long ago, the main regular alternative was an agitating table, in spite of its low effectiveness. Turning gadgets are likewise utilized as a part of an extremely set number of units; for example, they are used at Nevada, Marvel Loch, Western Australia, and Round Mountain. Serious cyanidation, despite its higher recuperations, always accomplishes a high level of acknowledgment, perhaps in view of the absence of a business plant.

Different aspects – for example, security issues, poor operability, and upkeep of these circuits joined with fast modified and the tastefulness of the carbon-in-pulp (CIP) and carbon-in-filter (CIL) process – fit for accomplishing high recuperations, and saw a decrease in dependence on gravity as an essential method for focus. It was enhanced by moving towards improved, less expensive units while keeping an eye on levels and mechanized procedures. This control of working price makes conceivable the operation of lower review metals. It is exemplified by second rate oxide orebodies of Western Australia, numerous with head reviews about 1 g/t. In any case, some metal bodies have been found to have properties which don't lend themselves to higher recuperation via the immediate cyanidation course. Coarse free gold and the gold related with complex minerals have a tendency to muddle the cyanidation procedure. Coarse gold builds the home time needed to accomplish higher recuperations by cyanidation. These issues are for the most part intensified for the coarse pounds typically connected with second-rate metals. A superior comprension of these issues and the improvement of bigger, more dependable gravity units, and also concentrated cyanidation, have proclaimed the arrival of gravity recuperation.

1.4.1.2 Froth Floatation

A large portion of the major work on the buoyancy of gold has been done utilizing high-immaculateness gold and gold–silver amalgams with the motivation b›ind deciding collector–gold communications and the idea of adsorption of gatherer particles or atoms onto the gold surface. What's more, some work has been done to determine whether or not unadulterated gold has a characteristic hydrophobicity and subsequently some level of common floatability. The buoyancy qualities of gold or gold minerals found in refractory sulphide and copper metals have not been depicted in detail in the writing. The meagre appropriation of discrete gold minerals and particles, and their exceedingly low focus in metals, are the foremost explanations b›ind the absence of basic work on gold buoyancy. A lot of work has been accounted for on particular metals, yet such investigations once in a while recognize the buoyancy of local gold and other gold minerals. Buoyancy of gold metals covers a wide field and it is a fairly troublesome subject to sum up. Most problems in gold mineral buoyancy are not associated with gliding metallic gold. The buoyancy recuperation of free gold (throughout the text free gold is synonymous with freed gold) is generally influenced by physical requirements; for example, the shape and size of the gold particles and the strength of the foam. Liberated gold greater than around 150 mm buoys promptly with most authorities and specifically with xanthates and dithiophosphates. At the point when free gold is coasted with other sulphide minerals the degree of air pocket stacking of sulphide particles may give an obstruction towards the attachment of free gold, consequently decreasing buoyancy execution. As of not too long ago, research investigations have commonly centred around the individual buoyancy conduct of every gold-bearing mineral in engineered blends and not on blends of sulphide minerals in "genuine" metals (Teague et al., 1999a). In the buoyancy procedure, the primary concoction impacts are reagent sort and mash pH. As of late, there has been a need to work circuits at direct pH levels, to enhance detachment efficiencies while treating complex poor-quality minerals, to lessen expenses of reagents, to create reagents that are steady finished a wide pH go, and to exploit the synergistic advantage of blend authority frameworks. This has prompted a regularly expanding examination to grow new authorities and blends for the buoyancy of gold-bearing minerals (Nagaraj, 1994, 1997).

1.4.1.2.1 Frothers in Gold Flotation

The quality and strength of the foam is essential when gliding free gold. There is an impression of there being an inclination for polyglycol ether-construct frothers in light of most Collector structure – singular authorities Cytec reagent subtle elements sodiumisobutyl xanthate AERO 317, potassium amyl xanthate AERO 343, xanthogenformate detailing AERO

3758, diisobutyldithiophosphate AERO 3477, mercaptobenzothiazole AERO 404, monothiophosphate AERO 6697, dithiophosphate definition Reagent S-9810, collector sythesis – mixed gatherers, dithiophosphate/monothiophosphate plan AERO 7249, monothiophosphate/dithiophosphate plan AERO 8761, dithiophosphate/monothiophosphate/dithiophosphinate, definition Reagent S-9913, dithiophosphate/mercaptobenzothiazole detailing AERO 405, dithiophosphate/mercaptobenzothiazole definition AERO 7156, thionocarbamate/dithiophosphate plan AERO 3926, thionocarbamate/dithiophosphate detailing AERO 473, modified thionocarbamate/dithiophosphate details AERO 5744/5, dicresyldithiophosphate plan reagent S-8985, modified thionocarbamate/dithiophosphate/monothiophosphate detailing reagent S-9889, ethyl octyl sulphide definition reagent S-701, and dithiocarbamate/ sodium hydrosulfide detailing reagent S-3730 are some of the well-known reagents to be used as frother in gold flotation. At the point when selectivity is required or, on account of copper–gold metals, where a copper think is sold to a smelter, a weaker frother (for example, methyl isobutyl carbinol (MIBC)) is favoured. The decision of a molecule measure adjusted frother additionally is a vital thought in gold buoyancy as this advances composite molecule recuperation in the forager buoyancy circuit. When in doubt, the glycol or polypropylene glycol methyl ether frothers are perfect for this application (Klimpel, 1997). The mixed inter froth frothers have discovered wide acknowledgment in the Australian gold industry (Goold, 1990).

1.4.1.2.2 *Activators in Gold Flotation*

Enactment infers enhanced floatability of a mineral after the expansion of a solvent base metal salt or sulfidizer. It is by and large felt that the metal or sulphide particle adsorbs onto the mineral surface in this manner changing its surface chemical properties. Thus, the buoyancy reaction can be enhanced or potentially the pH scope of buoyancy for the mineral can be expanded, the rates of buoyancy expanded and selectivity moved forward. It is broadly acknowledged that the principal reason for copper sulphate in the buoyancy of sulphide gold transporters is to improve the buoyancy of the sulphides and, specifically, pyrrhotite (Mitrofanov and Kushnikova, 1959), Arsenopyrite (Gegg, 1949; O'Connor et al., 1990), and pyrite (Bushell and Krauss, 1962). The succession of copper sulphate expansion (before or after the authority) is critical (Teague et al., 1999b; Monte et al., 2002). The expansion of copper sulphate has been found to build the rate of pyrite buoyancy, giving a general increment in gold recuperation as a result of the gold relationship with pyrite (Allison et al., 1982; Duchen and Carter, 1986). In this application, the actuation with copper sulphate improved the buoyancy of coarse pyrite. The adsorption of copper onto pyrite and pyrrhotite is R. Dunne 316 pH-needy, smaller amounts being adsorbed at antacid conditions. For a few metals, the expansion of copper sulphate at middle of the road pH esteems (for example, 7–10) might be destructive and may lessen pyrite recuperation (Bushell, 1970). The use of sulfidizers (sodium sulphide and sodium hydrosulfide) to upgrade the buoyancy of oxidized metals is outstanding (Jones and Woodcock, 1984; Oudenne and de Cuyper, 1986; O'Connor and Dunne, 1991). The principal nitty gritty research facility investigation of the impact of sodium sulphide on the buoyancy of gold-bearing minerals was attempted in the mid-1930s (Leaver and Woolf, 1935). The result from this investigation was that, all in all, sodium sulphide impedes the buoyancy of gold, despite the fact that for a few minerals there was an advantage in its expansion. Comparative remarks are to be found in the writing since that time (Taggart, 1945; Aksoy and Yarar, 1989). Sulphide particles seem to go about as flotation activators at low focuses (under 105 M) and as a solid depressant at fixations over 105 M (Aksoy and Yarar, 1989). The addition of sulphide particles changes over a few coatings on mineral surfaces in sulphides (Healy, 1984) and the resulting xanthate expansion will advance buoyancy. For fruitful enactment, the sulphide activator ought to be included gradually and at starvation amounts.

1.4.1.2.3 Depression of Gold in Flotation

The following are examples of depressing agents for local gold that are normally presented amid the buoyancy procedure incorporate mixes: calcium particles, chloride particles, calcium carbonate, cyanide, sodium silicate, sodium sulphite, ferric and substantial metal particles, tannin and related mixes, starch and other natural depressants, and numerous others (Taggart, 1945; Bro–man et al., 1987; Marsden and House, 1992; Lins and Adamian, 1993; Allan and Woodcock, 2001; Chryssoulis, 2001). These may aggressively adsorb on the gold surface, accordingly preventing the adsorption of the collector/s included. It has likewise been recommended that the ferric particles, which would be hydrated oxides, may go about as a physical hindrance between the air pocket and gold surface, yet this impact is reversed basically by washing with water (Aksoy and Yarar, 1989). Be that as it may, buoyancy of local gold frequently continues tastefully within the sight of a large number of these mixes. By and large, the outcomes revealed by various creators are not in great attention (Allan and Woodcock, 2001). It is likely that different parts in arrangement or on the surface of the gold that were not measured give the response to the distinctive results. Lime can't be considered as only a pH modifier and studies have demonstrated that calcium is emphatically adsorbed on sulphide minerals and gold at pH esteems ≥10 (Healy, 1984; Chryssoulis, 2001). This adsorption is improved if overabundance of sulphate in the mash advances calcium-sulphate coatings on particles. Desorption of calcium from the surface by lessening the pH can be helped by the utilization of particular calcium-complexing particles; for example, polyphosphate. It will advance more if the calcium discharge is endeavoured while including the overabundance activator; at that point a hydrophilic hydroxide covering can come about (Healy, 1984). Metal particles introduced from the circuit water, or from solvent metal particles in the mineral, may adsorb and nucleate as hydroxide coatings on all molecule surfaces, therefore inhibiting authority adsorption. The suggested strategy for buoyancy treatment (Healy, 1984) is to work at as low a pH esteem as common sense, stay away from fast increments in pH, include activator gradually or condition independently, and keep the tailings dam at a pH of least dissolvability (i.e., most extreme metal hydroxide precipitation). Commonly, the influential conditions for gold flotation can be understood as the following:

i. Eh of the buoyancy mash/floating
ii. Flotation gasses and the effect of oxidation on buoyancy
iii. Modification of pH for buoyancy
iv. Particle size and shape in buoyancy
v. Electrical twofold layer
vi. Flotation energy

1.5 Classification of Gold-Bearing Ores

By and large gold minerals can be delegated "free processing," "complex," or "obstinate." Free processing minerals give a gold recuperation >90% with an ordinary 20–30h cyanidation drain. Adequate cyanide is added to leave a convergence of 100–250 ppm at about pH 10, before the finish of the filter. Minerals that don't give monetary gold recuperation through customary cyanidation are named headstrong. (In this audit, metals that give worthy financial gold recuperation just with the utilization of essentially higher cyanide or oxygen prerequisites will be alluded to as unpredictable, while the expression "stubborn" will be utilized when upgraded reagent expansion still gives deficient gold recuperation.) Metals may not react to traditional cyanidation for three fundamental reasons. Right off the bat, in profoundly recalcitrant metals, the gold can be secured up in the mineral grid with the goal that filter reagents can't achieve it. Besides, in complex metals responsive minerals in the metal can expend the filter reagents inside responses, it might be lacking cyanide, and, additionally, oxygen in the mash to drain the gold. Third, parts of the mineral may adsorb or hasten the broken-down gold cyanide complex so it is lost from the filter alcohol. A few minerals may have commitments from each of these elements, which will impact the handling methodology. The classification of gold ores at a glance can be understood by Figure 1.3.

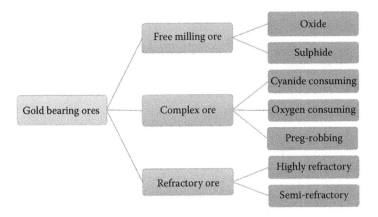

FIGURE 1.3
Classification of gold bearing ore-bodies.

1.5.1 Free Milling

Placer minerals do not require any pretreatment and are traditionally recuperated by physical strategies, like gravity treatment. Present day gravity treatment units, for example, concentrators (Knelson and Mathieson, 1991; McAlister, 1992), are being used for gold recuperation and supplement gravitational constrain with radial drive to empower the catching of better gold particles. The processing of free milling ores is shown in Figure 1.4.

Another physical procedure is coal-oil agglomeration (Bellamy et al., 1989; Cadzow and Lamb, 1989) with application mainly to the alluvial minerals. For substance recuperation courses pretreatment with the expectation of complimentary processing minerals is normally constrained to pulverizing and crushing, opening gold for cyanidation, essentially without being completely freed. The drain rate and extreme gold recuperation differ to some degree with the fineness of pound. Second-rate metals will probably be dealt with by stack draining (Dorey et al., 1988; McLean, 1988). The gold should be open to cyanide; consequently, the penetrability of the metal, in addition to other things, will manage the level of mineral arrangement required. The most minimal preparing expense will accompany run-of-mine dump draining. Less penetrable minerals require smashing before load filtering, maybe while fine particles ought to be immobilized by agglomeration with lime, concrete (O'Brien, 1982; McClelland and Van Zyl, 1988), or polymeric reagents (Polizzotti and Robertson, 1992) to anticipate permeation issues. The lower capital and working expenses of stack filtering are to some degree balanced by bring down gold extraction contrasted and ordinary processing (Potter, 1981), with commonplace store drain recuperations of 60%–80%.

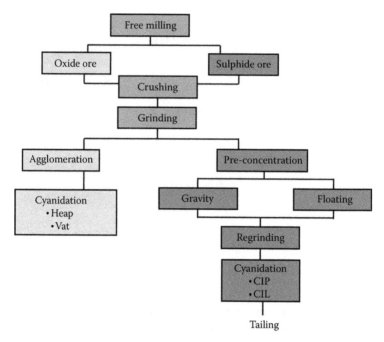

FIGURE 1.4
A typical processing of free milling gold bearing ores.

1.5.2 Complex Ores

The complex gold-bearing ores might be classified according to their high cyanide consumption, being highly oxygen demanding, and by the conduct of gold's preg-robbing, and processed according to their individual classification as shown in Figure 1.5.

1.5.2.1 Cyanide Consuming Complex Ore

The events of gold in electrum can slow gold draining energy and expanded cyanide utilization. More genuine cyanide utilization can emerge from the response of a few oxides or sulphides. These side-responses cause over the top reagent utilization which expands the cost of generation and in addition, diminishes gold recuperation. Complex metals, for the most part, have scope for sulphide crystals (for example marcasite, pyrrhotite, covellite, digenite, chalcocite, antimony and arsenic sulphides, and sulphides of zinc). Elective procedures, for example, the CuTech procedure (Sceresini and Staunton, 1991), proposed cyanide and copper recuperation, yet have not been utilized financially. Vitrokele TM route had likewise been designed especially for cyanide recuperation through stacking cyanide of copper buildings commencing desolate drained mash on uncommon particle trade pitches (Clarke, 1991) trailed by cyanide and some further gold recuperation. On the other hand,

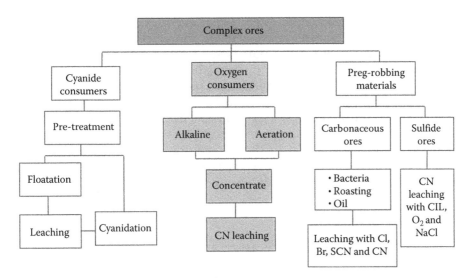

FIGURE 1.5
Processing of complex gold bearing ores as their classification.

particle buoyancy (Mudder and Goldstone, 1989; Engel et al., 1991) or particle trade gums (Mackenzie, 1991) had proposed on balf of a particular gold recuperation within sight for cyanide of copper buildings. By and large the most sparing methodology for sulphide mineral had created copper and gold buoyancy, on balf of metal recuperation through purifying (Johns and Green, 1991). Sometimes, sustained particles might potentially profit through an organized form of sulfidization (Jones et al., 1988; Englbardt, 1990; Dunne, 1991) that could deliver a worthy metal review aimed at smelters. Smelling cyanide salt (Hunt, 1901; Jarman and Le Gay Brereton, 1905; Lewis, 1990), Br (Putnam, 1950), thiourea (S›ic, 1988), and thiosulfate (Bilston et al., 1983; Zipprain et al., 1988) were entirely recommended as option components of gold for disintegration within sight of reserves of copper. Thiosulfate and bromine contained nearly constrained impending designs for specific gold draining above copper. A plant located in Mauritania right now utilizes a drain framework. The option method was utilizing pre-leaching electrolytic form; for example, Intec procedure (Costello et al., 1992), for evacuating reserves of copper preceding CIP/CIL cyanidation. The Rothsay mine in Western Australia in 1990 attempted different choices (Everett and Moyes, 1992).

1.5.2.2 Oxygen Consuming Complex Ore

Receptive sulphides, like pyrrhotite, can prompt higher air prerequisites to convert ferrous iron to ferric besides the oxidation of sulphide to sulphate, which can be supplied through expansion oxidants, unadulterated air (Hoeckar and Watson, 1992), peroxide of hydrogen, or calcium (Brooy and

Komosa, 1992; Loroesch et al., 1989). Expansion of oxidants could be organized through broken up air terminal in the primary drain container (Brooy and Komosa, 1992) yet nearby could issues a through pattern adjustment, especially through brine water (processed). Sulphides likewise respond through cyanide that produces thiocyanate (Loroesch et al., 1989; Brooy and Komosa, 1992; Monhemius, 1992), handling choices delineated. If the receptive sulphides are fruitless, at that point specific buoyancy after air circulation can be utilized to dismiss the effectively oxidized sulphides in the buoy appendage. Then again, by the side of bring down levels of pyrrhotite soluble before aeration might be adequate for passivating pyrrhotite through hydroxide or oxide films (Luthy and Bruce, 1979).

1.5.2.3 Preg-Robbing Ore

Preg-robbing conduct may emerge from carbonaceous materials via adsorbing the gold onto the carbon surfaces. Utmost well-known illustrations are Carlin Metals in the United States wherever carbon containing particles debasement ringlets $Au(CN)_2$ (Brooy, 1993). Carlin peroxidation of metal through chlorination neutralizes the material containing carbon (Osseo-Assare et al., 1984). On the other hand, broiling/pretreatment through bacteria (Brunk et al., 1988) could devastate preg-robbing conduct and gold could recuperate through customary lixiviant draining. The preg-robbing effect can be lessened through diesel extension/lamp fuel for the removal of carbonaceous issue in mineral. Nonetheless, because technique influences adsorption of gold cyanide onto carbon (CIP/CIL), in this way it warrants wary solicitation. Carbon containing materials in minerals do not have an indistinguishable gold stacking limit from enacted lower surface carbon region that might be up to 40 g/t gold (Harris, 1992a,b).

1.5.3 Refractory Ores and Pretreatments

The refractoriness of gold bearing ore further can be classified into four types, as per the gold dissolution efficiency under a standard cyanidation condition, and is shown in Figure 1.6. In a gold-bearing refractory ore, the ultrafine particles of gold remain disseminated throughout the minerals and resistant to exploitation by the standard operating practices of gold processing.

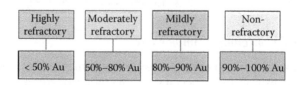

FIGURE 1.6
The classification of ore refractoriness with respect to the gold dissolution efficiency.

They often require a pretreatment in order to achieve an effective recovery of the precious metal. Commonly, the refractory ore includes sulphide minerals and/or organic carbon. The sulphide minerals are impermeable with occlude gold particles, which hinders the complexation of gold with lixiviant used; whereas, organic carbon may cause adsorption loss of the dissolved gold in a similar manner with the adsorption onto activated carbon. Notably, the preg-robbing carbon is washable due to it being significantly finer.

Various kinds of pretreatment options can be employed (as shown in Figure 1.7) for processing the refractory ores; however, only the major applications are discussed in detail.

1.5.3.1 Roasting

Roasting is a common pre(heat-)treatment process employed to oxidize both the sulphur and organic carbon in the presence of air and/or oxygen. The process converts the sulphur contents to gaseous hazards; toxic products are the major disadvantage. Depending on the ore types, roasting can be a single or two stages. The first method belongs to direct heat treatment in the presence of oxidizing atmosphere. The latter process consists of the conversion to a porous intermediate followed by a complete oxidation in the presence of oxidizing atmosphere. The sulfation or chlorination roasting is also performed in limited options. Commonly, the roasting of metal sulphides can be written as:

$$2MS + 3O_2 = 2MO + 2SO_2 \tag{1.1}$$

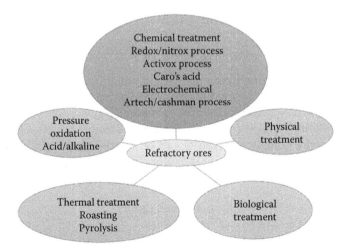

FIGURE 1.7
Possible pre-treatment procedures for processing the refractory gold ores for achieving an efficient recovery of gold in leaching operation.

Where M = Zn, Cu, or Pb. However, the conversion of roasted products of iron sulphide is not so simple. The stepwise reactions (one or more at once) may take place as:

$$3FeS_2 + 8O_2 = Fe_3O_4 + 6SO_2 \tag{1.2}$$

$$3FeS + 5O_2 = Fe_3O_4 + 3SO_2 \tag{1.3}$$

$$Fe_3O_4 + O_2 = 6Fe_2O_3 \tag{1.4}$$

In variance with the Equations 1.2–1.4, pyrite decomposes to pyrrhotite and volatilized sulphur as intermediates which further oxidizes to SO_2 in the presence of oxygen as:

$$FeS_2 = FeS + S \tag{1.5}$$

$$S + O_2 = SO_2 \tag{1.6}$$

Also, in the case of arsenopyrites, several reactions are supposed to take place as follows:

$$12FeAsS + 29O_2 = 4Fe_3O_4 + 6As_2O_3 + 12SO_2 \tag{1.7}$$

$$FeAsS = FeS + As \tag{1.8}$$

$$4As + 3O_2 = 2As_2O_3 \tag{1.9}$$

$$As_2O_3 + O_2 = As_2O_5 \tag{1.10}$$

$$Fe_2O_3 + As_2O_5 = 2FeAsO_4 \tag{1.11}$$

It is imperative to mention that most of the refractory gold is associated with iron and arsenic; hence, the oxidation of entire sulphides is primarily important. The unconverted sulphides may consume large amounts of cyanide in comparison to the reagent consumed by the oxidized products. On the other side, humic acid, coal, and graphite carbon as the source of carbonaceous substance must be oxidized (as in the following reaction) prior to leaching to prevent the preg-robbing.

$$C + O_2 = CO_2 \tag{1.12}$$

Notably, the oxidation rate of some carbonaceous matter can be slower than the sulphide minerals, as a high evolution of CO_2 at the initial stage may form film that can hinder the further oxidation of the mineral surface. The roasting temperatures, gas phase composition, and particle size of minerals are some of the influential factors for efficient roasting yield. In some of the cases pyrolysis of the refractory minerals has also been studied for the same purpose (Dry and Coetzee, 1986; Graham et al., 1992).

1.5.3.2 Biological Pretreatment

Pretreatment by microbial activity of refractory gold-bearing minerals was the subsequent application area after the dump leaching of low-grade copper ores started in ~1900. Since then, after the passing of a century of commercial application of the bio-oxidation, it has been retrofit to the ordinary cyanidation followed by CIP/CIL process (Bruynesteyn, 1988). In Harbor Beams gold pithead at Western Australia, 40 t/day arsenopyritic ponders are handled utilizing the Gencor BIOX technology (Barter et al., 1992), while Facilitate Arctic at Wiluna handled a 115 t/day pyritic/arsenopyritic ores (Odd and Baxter, 1993).

Usually the mesophile microorganisms like *thiobacillus thio-oxidans* and *thiobasillus ferro-oxidans* are employed to treat the sulphide ores at ambient (~35°C) temperatures. However, the use of moderate thermophiles (*sulfobacillus acidophiles*) and thermophiles (*sulfolobus*) is increasing to achieve faster oxidation at commercial operations due to their possible survival at elevated temperatures. Microorganisms derive the required energy from the oxidation of iron and sulphur species, but the requirements of oxygen, nitrogen, and carbon must be supplied either from outside the system or from the ore bodies. The microbial catalysed oxidation of arsenopyrite can be written as:

$$4FeAsS + 13O_2 + 6H_2O = 4H_3AsO_4 + 4FeSO_4 \qquad (1.13)$$

$$2FeAsS + 7O_2 + H_2SO_4 + 2H_2O = Fe_2(SO_4)_3 + 2H_3AsO_4 \qquad (1.14)$$

$$2FeAsS + 6O_2 + Fe_2(SO_4)_3 + 4H_2O = 4FeSO_4 + H_2SO_4 + 2H_3AsO_4 \qquad (1.15)$$

Notably, a controlled partial oxidation can be achieved only in the bio-oxidation process that is not possible with either of the oxidation processes. It ultimately results in a lower requirement of heat and oxygen to the system, producing less acid with a corresponding lesser need of neutralization.

1.5.3.3 Pressure Oxidation Pretreatment

In spite of the requirement of a high capital cost, several commercial plants are using the pressure oxidation treatment prior to undergoing the cyanidation of refractory minerals (Litz and Carter, 1988; Linge, 1992). The reactions are performed in autoclave to withstand high pressure and temperatures. In general, the refractory ore containing sulphur content higher than 4% can be treated by autoclaving to efficiently liberate the gold in the cyanidation process. The procedure is most appropriate for minerals that require a sulphide decimation finish.

1.5.3.3.1 Acid Pressure Oxidation

Sulphide minerals can rapidly decompose in acidic media at elevated pressure and temperature, utilizing oxygen as the prime source of oxidation along with the ferric ions present in the system. The rate of oxidation primarily depends on the mass transfer rate of oxygen to the mineral surface, slurry temperature, the way of oxygen supply, and agitation speed. Typically, 150–700 kPa pO_2 are applied with a pressure of 1500–3200 kPa. The stoichiometric requirement of 20 kg oxygen for one-ton sulphide mineral has been observed; however, the oxygen utilization efficiency may vary from 50% to 90%. In a 4–5 compartment autoclave, the run of the mill working conditions are: temperature 170°C–225°C (to ensure the irreversible oxidation of sulphur) for 1–3 h (Berezowsky et al., 1991). In the autoclave, pyrite and arsenopyrite at first broken up to frame ferric, sulphate, and arsenate particles which are changed by hydrolysis into solids; for example, scorodite, haematite, fundamental iron sulphates, and jarosites (Berezowsky and Weir, 1989; Berezowsky et al., 1991).

The major oxidation reactions at pH < 2 under the elevated atmosphere can be written as:

$$2FeS_2 + O_2 + 4H^+ = 2Fe^{2+} + 4S + 2H_2O \tag{1.16}$$

$$2Fe_7S_8 + 7O_2 + 28H^+ = 14Fe^{2+} + 16S + 14H_2O \tag{1.17}$$

$$4FeAsS + 5O_2 + 8H^+ = 4Fe^{2+} + 4HAsO_2 + 4S + 2H_2O \tag{1.18}$$

$$4CuFeS_2 + 3O_2 + 12H^+ = 4Cu^+ + 4Fe^{2+} + 8S + 6H_2O \tag{1.19}$$

Additionally, the ferrous ions oxidize as:

$$4Fe^{2+} + O_2 + 4H^+ = 4Fe^{3+} + 2H_2O \tag{1.20}$$

The sulphur generated in the previous reactions can agglomerate to surface passivation of the mineral, preventing the gold from being amenable to leach. It can be avoided at a higher temperature (>170°C) by irreversible conversion of sulphur to sulphate ion, as follows:

$$2S + 3O_2 + 2H_2O = 4H^+ + 2SO_4^{2-} \tag{1.21}$$

The carbonates, if any, react with the sulfuric acid to precipitate gypsum as:

$$CaCO_3 + H_2SO_4 = CaSO_4 + CO_2 + H_2O \tag{1.22}$$

Among all of the oxidation processes, the acid pressure oxidation yields the highest gold recovery (90%–95%). However, some of the carbonaceous ores did not respond well due to the adsorption loss of dissolved gold and formation of CO_2 (as shown in Equation 1.22) that can cause it to lower the efficiency of oxidation by itself consuming the oxygen of the system.

1.5.3.3.2 Alkaline Pressure Oxidation

Similar to other conditions such as the acidic pressure oxidation under elevated temperature, the oxidation performed only differs at pH under neutral or slightly alkaline range. This is suitable for carbonaceous ores (>10% carbonate) that contain low sulphur. No addition of acid is required while the acid generated in-situ is neutralized by carbonate in the feed material. One of the advantages of this process is the formation of silver jarosite is not possible at the operated pH range. Hence the recovery of coexisting silver would be higher; however, a little gold may also be leached by complexing with *in*-situ formation of thiosulfate species. Notably, a higher pressure is necessary to maintain the desired pO_2, because the large amount of CO_2 evolved by high carbonate contents in autoclave may dilute the dissolved oxygen therein. The major oxidation reactions under a mild alkaline solution and at an elevated atmosphere can be written as:

$$4FeS_2 + 15O_2 + 14H_2O = 4Fe(OH)_3 + 16H^+ + 8SO_4^{2-} \quad (1.23)$$

$$4Fe_7S_8 + 69O_2 + 74H_2O = 28Fe(OH)_3 + 64H^+ + 32SO_4^{2-} \quad (1.24)$$

$$2FeAsS + 7O_2 + 8H_2O = 2Fe(OH)_3 + 2H_3AsO_4 + 4H^+ + 2SO_4^{2-} \quad (1.25)$$

$$4CuFeS_2 + 17O_2 + 18H_2O = 4Cu(OH)_2 + 4Fe(OH)_3 + 16H^+ + 8SO_4^{2-} \quad (1.26)$$

1.6 Extraction Techniques of Gold Ores

1.6.1 Amalgamation

An amalgam is simply an alloy of mercury with gold, and the process of amalgamation has been in practice for thousands of years to recover the precious metal of finer size. Mercury is basically used as a collector metal bringing free gold particles in contact to literally dissolve gold into it. Thus, the collected amalgam is subjected to separate the two metals for the recovery of gold and recycling of mercury. There are numerous ways to contact the mercury with gold; however, it needs a proper mixing of them in a clean form. Any kind of coating, especially one that is oily in nature, prevents the proper mixing of metals. In most of the old stamp mills, a metal plate surface-coated with a thin layer of mercury was used as an amalgamator in which the gold bearing ores/concentrates were charged slowly over the plate, adhering gold to the mercury. In such an open system of amalgamation, any mercury that went out with the tailings could endanger the environment due to the poisonous characteristics of mercury. The closed system is therefore better to use. Recently, the use of charged mercury (which is done by introducing a small quantity of active metal like sodium or potassium) is practiced to aggressively amalgamate the gold.

1.6.2 Chlorination

Prior to introducing cyanidation to gold metallurgy, chlorination was extensively used especially to the refractory ores that were difficult to leach by gravity concentration and amalgamation. Due to this, chlorination was also employed in pretreatment of sulphide and carbonaceous ores, dissolving a significant amount of gold in the solution. However, the formation of chloride complex of gold was known way before the 17th century with aqua regia leaching of gold. The chlorination process will be described separately in the chapter that discusses the halide leaching of gold.

1.6.3 Cyanidation

Despite being highly toxic, the cyanide leaching of gold is the most common technique in the exploitation of gold from its (high and low grade) ore bodies. Vat leaching is used for high grade ore processing, in which the ore slurry is treated under agitation in large tanks containing the solution of sodium cyanide. For low grade ores, heap leaching is practiced in which a dilute solution of sodium cyanide is sprayed from the top of the heaped pile. The cyanide solution percolates down with time and leaches the gold, and the gold bearing solution can be collected from the bottom of the heap. During cyanidation, the metallic gold is oxidized to leach in a dilute alkaline solution of cyanide. The cyanidation reaction of gold was originally given by Elsner in 1849; however, numerous studies on leaching mechanism and parametric influences have been carried out until now. These will be discussed separately in the cyanidation chapter.

1.7 Gold Metallurgy and Environmental Impact

Despite the shiny, ruddy yellow colour, gold is termed as *"dirty"* gold due to the process requirement of huge toxic chemicals and the generation of tons of wastes that is difficult to grasp. As estimated, an equivalent weight of the

Eiffel tower is generated every 42 s by the gold mining operation (http://www.theworldcounts.com). Due to stringent environmental rules and regulations with more and more concerns regarding public health and the aquatic system, there is immense pressure on the gold mining and metallurgy industry to reduce its environmental footprint. Although more sustainable process and technologies are being developed, there needs to be more information about their direct and indirect impact on the environment during different processing steps.

Gold is commonly mined in enormous open pits. The creation of new mines disrupts the local ecosystem, adversely affecting the interdependent relation between the elements of a healthy ecosystem including soil, water, plants, and animals. Dusts that are generated from open mines get mixed with air and can cause breathing issues. The study reveals that the mine sites' water, soil, and plants are contaminated by more than a dozen of heavy metals including chromium, vanadium, manganese, copper, nickel, cadmium, lead, cobalt, aluminium, zinc, strontium, barium, and iron (Abdul-Wahab and Marikar, 2012). A high concentration of aluminium and iron in well-water increases heavy metals uptake by plants and the blue colour of surface water is due to high copper contamination indicated for a significant immobilization of heavy metals. Tailing dumped near water supplies causes a loss of up to 80% of aquatic life. The high temperature operations like roasting of sulphides ores or calcination can release gaseous pollutants into the air. A mine of high grade ore that might contain a few hundredths of an ounce gold per ton, not only requires high energy but may also lead to toxic mine drainage. And the processing of acid generating sulphide ores generates sulfuric acid. In the extraction process, the uses of hazards like mercury in gold amalgamation and cyanide as the most suitable lixiviant are the biggest threats to the environment. In the recovery and refining process using smelter, a high gaseous emission causes breathing problems.

A study on life cycle assessment for estimating the uses of energy, water, waste generation, and emissions of greenhouse gases can clear the picture on the impact of gold metallurgy (including mining). For this, consider the entire processing operations: from open pit mining to ore beneficiation, cyanide leaching to carbon adsorption, and chlorination to gold refinement are a necessity (Norgate and Haque, 2012). For 3.5 g gold per ton of ore, the production of one-ton gold (from non-refractory ores) approximately consumes 200,000 GJ energy and 260,000 t of water, producing 18,000 t of GHGs and 1,270,000 t of solid wastes. The advent of hydraulic mining needed massive amounts of water, which allowed miners to go to deeper places for gold deposit than the simpler panning of gold. Moreover, hydraulic mining could displace the earth ~4000 cubic yards per day, enough for flattening many hills and mountains. The emissions and energy use increase ~50% higher while processing the gold-bearing refractory ores. Nevertheless, the production quantity of gold is much less than steel or aluminium; the environmental footprint for worldwide gold production is also less despite a higher magnitude of negative impact on the environment. Notably, the mining, crushing, and grinding of gold bearing ores contribute more to the environmental impact; the processes are highly dependent on the content of gold. A lower grade ore would require more comminution and consumes more electricity. With a common decline in grade of ores, or say depleting gold, the need to handle the environmental impacts of gold processing (mining and metallurgy) is becoming more important.

References

Abdul-Wahab, S.A., Marikar, F.A. 2012. The environmental impact of gold mines: Pollution by heavy metals. *Central European Journal of Engineering*. 2(2): 304–313. DOI:10.2478/s13531-011-0052-3.

Aksoy, B.S., Yarar, B. 1989. Natural hydrophobicity of native gold flakes and their flotation under different conditions. In: Dobby, G.S., Rao, S.R. (Eds), *Processing of Complex Ores*. Pergamon Press, New York, pp. 19–27.

Allan, G.C., Woodcock, J.T. 2001. A review of the flotation of native gold and electrum. *Minerals Engineering*. 14(9): 931–962.

Allison, S.A., Dunne, R.C., De Waal, S.A. 1982. The flotation of gold and pyrite from South African gold-mine residues. In: *14th International Mineral Processing Congress*, Toronto, ON, paper II-9.1.

Barter, J., Carter, J., Holder, N.H.M., Miller, D.M., Van Aswegen, P.C. 1992. Design and commissioning of a 40 tonne/day flotation concentrate biooxidation treatment plant at the Harbour Lights mine. In: *International Conference on Extractive Metallurgy of Gold and Base Metals*, Kalgoorlie, WA, 26–28 October, Aus.I.M.M, Melbourne, VIC, p. 113.

Bellamy, S.R., House, C.I., Veal, C.J. 1989. Recovery of fine gold from a placer ore by coal gold agglomeration. In: *Gold Forum on Technology and Practices 'World Gold '89'*, S.M.E., Littleton, CO, p. 347.

Berezowsky, R.M.G.S., Weir, D.R. 1989. Refractory gold: The role of pressure oxidation. In: *Gold Forum on Technology and Practices, 'World Gold '89'*, AIME, Littleton, CO, p. 295.

Berezowsky, R.M.G.S., Collins, M.J., Kerfoot, D.G.E., Torres, N. 1991. The commercial status of pressure leaching technology. *Journal of Metals*. 43: 9–15.

Bilston, D.W., Brooy, L.S.R., Woodcock, J.T. 1983. Use of thiourea leaching for gold and silver recovery. Status report AMIRA project 80/P127. Australian Mineral Industries Research Association, Melbourne, VIC.

Bro–man, B.R., Carter, L.A.E., Dunne, R.C. 1987. Flotation. In: *The Extractive Metallurgy of Gold in South Africa*, Monograph series M7, vol. 1. South African Institute of Mining and Metallurgy, Johannesburg, South Africa, pp. 235–275.

Brooy, L.S.R. 1993. Gold technology for complex and refractory ores. Final report AMIRA Project 89/P314. Australian Mineral Industries Research Association, Melbourne, VIC, p. 49.

Brooy, L.S.R., Komosa, T. 1992. Oxidant addition during gold ore processing. In: *International Conference on Extractive Metallurgy of Gold and Base Metals*, Kalgoorlie, WA, 26–28 October, Aus.I.M.M., Melbourne, p. 147.

Brunk, K.A., Ramadorai, G., Seymour, D., Traczyk, P.F. 1988. Flash chlorination—A new process for the treatment of refractory gold sulphides and carbonaceous gold ores. In: *Ratwlol Perth International Gold Conference*, Randol International, Golden, CO, p. 127.

Bruynesteyn, A. 1988. Biotechnology for gold ores: The state of the art. In: *Randol Perth International Gold Conference*, Randol International, Golden, CO, p. 141.

Burt, R.O., Mills, C. 1984. *Gravity Concentration Technology*. Elsevier, Amsterdam, Oxford, New York, Tokyo, p. 467.

Bushell, C.H.G., Krauss, G.J. 1962. Copper activation of pyrite. *Canadian Mining and Metallurgical Bulletin*. 55: 314–318.

Bushell, L.A. 1970. The flotation plants of the Anglo-Transvaal Group. *Journal of the Southern African Institute of Mining and Metallurgy*. 70: 213–218.

Cadzow, M., Lamb, R. 1989. Carbad gold recovery. In: *Gold Forum on Technology and Practices 'World Gold '89'*, S.M.E., Littleton, CO, p. 375.

Chryssoulis, S.L. 2001. Using mineralogy to optimize gold recovery by flotation. *Journal of the Minerals, Metals and Materials Society*. 53(12): 48–50.

Clarke, S. 1991. The Cutech process. In: *AMMTEC Processing of Gold-Copper Ores Colloquium*, AMMTEC, Perth, WA.

Costello, M.C., Ritchie, I.C., Lund, D.J. 1992. Use of ammonia cyanide leach system for gold copper ores with reference to the retreatment of Torco tailings. *Minerals Engineering*. 5(10–12): 1421.

Dorey, R., Zyl, V.D. Kiel, J. 1988. Overview of heap leaching technology. In: *Introduction to Evaluation, Design and Operation of Precious Metal Heap Leaching Projects*, S.M.E., Littleton, CO, p. 3.

Dry, M.J., Coetzee, C.F.B. 1986. The recovery of gold from refractory ores. In: *Gold 100, Proceedings of the International Conference on Gold. Extractive Metallurgy of Gold*, vol. 2. South African Institute of Mining and Metallurgy, Johannesburg, p. 259.

Duchen, R.B., Carter, L.A.E. 1986. An investigation into the effects of various flotation parameters on the flotation b›aviour of pyrite, gold and uranium contained in Witwatersrand type ores, and their practical exploitation. In: *Gold 100, Proceedings of the International Conference on Gold*, vol. 2. South African Institute of Mining and Metallurgy, Johannesburg.

Dunne, R. 1991. Auriferous sulphide flotation in Australia. In: *Randol Gold Forum Cairns'91*. Randol International, Golden, CO, p. 239.

Elvey, L.E., Woodcock, J.T. 1966. *The Australian Mining Metallurgical, and Mineral Industry*, vol. 3. Thompson Publications Australia Pty Ltd, Melbourne, VIC.

Engel, M.D., Leahy, G.J., Moxon, N.T., Nicol, S.K. 1991. Selective ion flotation of gold from alkaline cyanide solutions. In: *World Gold'91. Gold Forum on Technology and Practice*, Aus.I.M.M., Melbourne, VIC, p. 121.

Engbardt, D. 1990. Telfer gold mine: Sulphide ore treatment circuit. In: *Randol Gold Forum Squaw Valley'90*, Randol International, Golden, CO, p. 67.

Everett, P.K., Moyes, A.J. 1992. The Intec copper process. In: *International Conference on Extractive Metallurgy of Gold and Base Metals*, Kalgoorlie, WA, 26–28 October, Aus.I.M.M., Melbourne, p. 287.

Gegg, R.C. 1949. Milling and roasting at MacLeod-Cockshutt. *Canadian Mining and Metallurgical Bulletin*. 42: 659–665.

Goold, L.A. 1990. *Private Communication*. Chemical and Mining Services, Sydney, NSW.

Graham, J., Fletcher, A.B., Walker, G.S., Aylemore, M. 1992. A pyrolysis approach to refractory gold processing. In: *International Conference on Extractive Metallurgy of Gold and Base Metals*, Kalgoorlie, WA, 26–28 October, Aus.I.M.M., Melbourne, p. 441.

Hamburger, M., Hereford, A., Simmons, W. 2010. *Volcanoes of the Eastern Sierra Nevada: Geology and Natural Heritage of the Long Valley Caldera*. Indiana University Department of Geological Sciences, Bloomington, IN.

Harris, L. 1992a. Newmont's refractory ore program. In: *Randol Gold Forum Vancouver'92*, Randol International, Golden, CO, p. 149.

Harris, L. 1992b. Newmont's refractory gold ore program. In: *Randol Gold Forum Vancouver'92*, Randol International, Golden, CO, p. 465.

Healy, T.W. 1984. Pulp chemistry, surface chemistry and flotation. In: *Principles of Mineral Flotation, Wark Symposium*, Series No. 40, Aus.I.M.M., Melbourne, VIC, pp. 43–56.

Hoecker, W., Watson, S. 1992. Oxygen e~thanced leaching—Case studies. In: *Randol Gold Forum Vancouver'92*, Randol International, Golden, CO, p. 456.

http://www.theworldcounts.com/counters/environmental_effect_of_mining/environmental_effects_of_gold_mining

Hunt, B. 1901. Process of precipitating and recovering precious metals from their solutions. U.S. Patent. 689:190.

Jarman, A., Brereton, E.L.G. 1905. Laboratory experiments on the use of ammonia and its compounds in the cyaniding cupriferous ores and tailings. *Transactions of the Institute of Mining and Metallurgy*. XIV: 289.

Johns, M.W., Green, B.R. 1991. The resin in pulp program at Mint–. In: *Randol Gold Forum Cairns'91*, Randol International, Golden, CO, p. 363.

Jones, M.H., Wong, K.Y., Woodcock, J.T. 1988. Controlled-potential sulphidization and rougher-cleaner flotation of an oxide-sulphide copper ore. In: *13th Common Wealth Mining and Metallurgical Congress*, Aus.I.M.M., Melbourne, VIC, p. 33.

Jones, M.H., Woodcock, J.T. 1984. Application of pulp chemistry to regulation of chemical environment in sulfide mineral flotation. In: *Principles of Mineral Flotation, Wark Symposium*, Series No. 40. Aus.I.M.M., Melbourne, VIC, pp. 147–174.

Kettell, B. 1982. *Gold*. Oxford University Press, Melbourne, VIC.

Klimpel, R.R. 1997. An approach to the flotation of complex gold ores containing some free gold and/or some gold associated with easily floatable sulfide minerals. In: *World Gold '97*, Aus.I.M.M, Melbourne, VIC, pp. 109–113.

Knelson, B., Mathieson, D. 1991. Benefits from the introduction of Knelson concentrators in gold plants in Australia. In: *Randol GoM Forum Cairns '91*, Randol International, Golden, CO, p. 245.

Leaver, E.S., Woolf, J.A. 1935. Flotation of gold, effect of sodium sulfide. U.S. Bureau of Mines Report of Investigation No. 3275, Progress Report-Metallurgical Division, 11 Studies on the Recovery of Gold and Silver, pp. 23–38.

Lewis, P.J. 1990. Treatment of oxidised and primary copper/gold ores at Red Dome, Queensland, Australia. In: *Randol Gold Forum Squaw Valley'90*, Randol International, Golden, CO, p. 59.

Linge, H.G. 1992. New technology to refractory gold ores. In: *International Conference on Extractive Metallurgy of Gold and Base Metals*, Kalgoorlie, WA, 26–28 October, Aus.I.M.M, Melbourne, VIC, p. 339.

Lins, P.J.D., Adamian, R. 1993. Some chemical aspects of gold particles flotation. In: *XVIII International Mineral Processing Congress*, Sydney, vol. 5, Aus.I.M.M, Melbourne, VIC, pp. 1119–1122.

Litz, J.E., Carter, R.W. 1988. Comparative economics of refractory gold ore treatment processes. In: *Randol Perth International Gold Conference*, Randol International, Golden, CO, p. 133.

Loroesch, J., Knorre, H., Griffiths, A. 1989. Developments in gold leaching using hydrogen peroxide. *Mining Engineering*. 41(9): 963–965.

Luthy, R.G., Bruce, S.G. 1991. Kinetics of reaction of cyanide and reduced sulphur species in aqueous solutions. *Journal of Environmental Science and Technology*. 13(12): 1481.

MacKenzie, J.M.W. 1991. Henkel gold recovery technology. In: *AMMTEC Processing of Gold-Copper Ores Colloquium*, AMMTEC, Perth, WA.

Marsden, J., House, I. 1992. *The Chemistry of Gold Extraction*. Ellis Horwood, London, p. 597.

McAlister, S. 1992. Case studies in the use of the Falcon gravity concentrator. In: *Randol GoM Forum Vancouver'92*, Randol International, Golden, CO, p. 131.

McClelland, G.E., Van Zyl, D. 1988. Ore preparation: Crushing and agglomeration. In: *Introduction to Evaluation, Design and Operation of Precious Metal Heap Leaching Projects*, S.M.E., Littleton, CO, p. 68.

McLean, E.J. 1988. When is heap leaching the best choice. In: *Economics and Practice of Heap Leaching in Gold Mining*, Aus.I.M.M., Melbourne, VIC, p. 5.

Mitrofanov, S.I., Kushnikova, J. 1959. Adsorption of butyl xanthate and Cu^{2+} ion by pyrrhotite. *Mine and Quarry Engineering*. 25: 362–364.

Monte, M.B.M., Dutra, A.J.B., Albuquerque, C.R.F., Tondo, L.A., Lins, F.F. 2002. The influence of the oxidation state of pyrite and arsenopyrite on the flotation of an auriferous sulfide ore. *Minerals Engineering*. 15: 1113–1120.

Mudder, T.I., Goldstone, A.J. 1989. The recovery of cyanide from slurries. In: *Randol Gold Forum Sacramento'89*, Randol International, Golden, CO, p. 111.

Nagaraj, D.R. 1994. A critical assessment of flotation reagents. In: Mulukutla, P.S. (Ed.), *Reagents for Better Metallurgy*. The Society for Mining, Metallurgy and Exploration Inc., Littleton, CO, pp. 81–90.

Nagaraj, D.R. 1997. Developments of new flotation chemicals. *Transactions of the Indian Institute of Metals*. 50(5): 355–363.

Norgate, T., Haque, N. 2012. Using life cycle assessment to evaluate some environmental impacts of gold production. *Journal of Cleaner Production*. 29–30: 53–63.

O'Brien, R.T. 1982. Agglomeration pre-treatment in heap leaching of gold and silver. In: *Carbon-in-Pulp Technology for the Extraction of Gold*, Aus.I.M.M., Melbourne, VIC, p. 297.

O'Connor, C.T., Bradshaw, D.J., Upton, A.E. 1990. The use of dithiophosphates and di-thiocarbamates for the flotation of arsenopyrite. *Minerals Engineering*. 3(5): 447–459.

O'Connor, C.T., Dunne, R.C. 1991. The practice of pyrite flotation in South Africa and Australia. *Minerals Engineering*. 4(7–11): 1057–1069.

Odd, P., Baxter, K. 1993. The Wiluna sulphide gold project: Feasibility to operation, Branch meeting, Aus.I.M.M., Perth, WA.

Osseo-Assare, K., Afenya, P.M., Abotsi, G.M.K. 1984. Carbonaceous matter in gold ores: Isolation, characterisation and adsorption b>aviour in aurocyanide solutions. In: Kudryk, V., Corrigan, D.A., Liang, W. (Eds), *Precious Metals: Mining, Extraction and Processing*, AIME, Warrendale, PA, p. 125.

Oudenne, P.D., de Cuyper, J. 1986. Reagents and flotation flow-sheet selection for the beneficiation of a complex sulfide ore containing copper and gold. In: *Proceedings of the 2nd International Symposium on Beneficiation and Agglomeration*, Bhubaneswar, India, pp. 358–364.

Petrovskaya, N.V. 1987. An outline of gold chemistry. In Boyle, R.W. (Ed.), *Gold: History and Genesis of Deposits*. Van Nostrand Reinhold, New York, p. 137.

Polizzotti, D.M., Robertson, J.J. 1992. Polymeric agglomerating agents for the gold mining industry. In: *Randol Gold Forum Vancouver'92*, Randol International, Golden, CO, p. 223.

Potter, G.M. 1981. Some developments in gold and silver metallurgy. In: *Extraction Metallurgy '81*, Institution of Mining and Metallurgy, London.

Putnam, G.L. 1950. Copper-ammonia process for cyanidation of complex ores. *Chemical Engineering and Mining Review*. 42: 347.

Richards, R.G. 1988. In: *4th International Gold Conference*, Rio de Janeiro, Brazil.

Sceresini, B., Staunton, W. 1991. Copper/cyanide in the treatment of high copper gold ores. In: *Fifth Aus I.M.M. Extractive Metallurgy Conference*, Perth, WA, 2–4 October, Aus.I.M.M., Melbourne, VIC, p. 123.

Seeger, P.A., Fowler, W.A., Clayton, D.D. 1965. Nucleosynthesis of heavy elements by neutron capture. *The Astrophysical Journal Supplement Series*. 11: 121.

Sₓic, O.A. 1988. An update on the K-process. In: *Randol Perth International Gold Conference*, Randol International, Golden, CO, p. 184.

Shoemaker, R.S. 1984. In: *Precious Metals, Conference Proceedings*. February 27–29.

Taggart, A.F. 1945. *Handbook of Mineral Dressing*, Section 12. John Wiley & Sons, New York, pp. 116–119.

Teague, A.J., Van Deventer, J.S.J., Swaminathan, C.I. 1999a. A conceptual model for gold flotation. *Minerals Engineering*. 12: 1001–1019.

Teague, A.J., Van Deventer, J.S.J., Swaminathan, C.I. 1999b. The effect of galvanic inter-action on the behaviour of free and refractory gold during froth flotation. *International Journal of Mineral Processing*. 57: 243–263.

Willbold, M., Elliott, T., Moorbath, S. 2011. The tungsten isotopic composition of the Earth's mantle before the terminal bombardment. *Nature*. 477(7363): 195–198.

Yannopoulos, J.C. 1991. *The Extractive Metallurgy of Gold*, 1st edition. Van Nostrand Reinhold, New York.

Zipprian, D., Raghavan, S., Wilson, J.P. 1988. Gold and silver extraction by ammoniacal thiosulphate leaching from a rhyolite ore. *Hydrometallurgy*. 19: 361–375.

2

Artisanal Gold Mining and Amalgamation

Sadia Ilyas[*] and Jae-chun Lee[†]

2.1 Introduction

The mining activities of gold conducted by an individual miner or small-scale enterprise with limited capital investment and production output can be termed as artisanal gold mining or ASGM (United Nations Environment Program, 2014). However, the International Labor Organization (ILO) has described artisanal mining as a labour-intensive activity that is carried out with a lack of mechanized facilities (Jennings, 1999), including the other characteristics of an informal sector designed to exploit the gold deposits in a non-scientific manner with a limited access to land and markets. These characteristics illustrate the cycle of poverty that exists around the communities located near to ASGM without any consideration of the issues related to health and environment, particularly where inefficient mining and processing techniques yield a small quantity of gold with low profit (Barry, 1996). In contrast to this, ASGM activities in many parts of the world are as important as the mining done on a large scale, particularly in terms of employment and their source of earning. ASGM plays a crucial role in rural and remote economies where no other source of income for the poor people is available, albeit while causing increased risks to their health and safety due to unsophisticated uses of mercury and other hazardous elements.

2.2 Process of Gold Extraction by ASGM

The unit processes used in gold extraction by ASGM are schematically shown in Figure 2.1 and briefly discussed here (Sousa et al., 2010; United Nations

[*] Mineral and Material Chemistry Lab, Department of Chemistry, University of Agriculture Faisalabad, Pakistan.
[†] Minerals Resources Research Division, Korea Institute of Geoscience and Mineral Resources, Daejeon, South Korea.

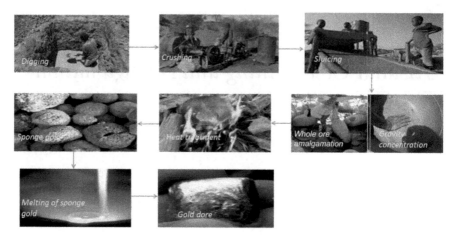

FIGURE 2.1
Schematic of unit processes used in gold extraction by ASGM.

Environment Program, 2015). Miners exploit alluvial deposits (river sediments) or hard rock deposits. Sediment or overburden is removed and the ore is mined by surface excavation, by tunnelling, or by dredging. For efficient liberation of gold, the required crushing and milling of hard rock are primarily done either manually by hammering or with machines, followed by the milling. In some cases, gold is further separated from other materials by concentration. Different methods and technologies (e.g. sluices, centrifuges, vibrating tables, etc.) may be used to concentrate the liberated gold. Due to the density difference of gold often being higher than other minerals in ore, many techniques utilize the gravity concentration. The concentrates are then subjected to amalgam with a large quantity of mercury. The obtained amalgam is heated to separate gold via vaporizing the mercury. In open burning practice, all of the mercury vapor emitted into the environment causes severe air pollution. The sponge gold obtained after evaporating the mercury is further heated to remove the residual mercury and other impurities, yielding pure gold.

2.3 Gold Amalgamation

Amalgamation is the most important unit operation of the ASGM, in which mercury in its elemental form is used to obtain the amalgam alloy with trapped gold. Historically, the use of mercury in gold metallurgy has been practiced for 4500 years (Malm, 1998), basically as a collector metal for gold. Thus, collected amalgam is subjected to separate the two metals for recovery of gold and recycling of mercury.

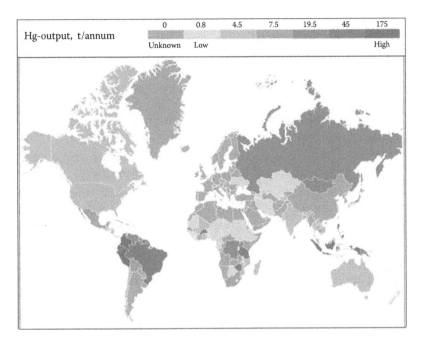

Hg-output, t/annum	0	0.8	4.5	7.5	19.5	45	175
	Unknown	Low					High

FIGURE 2.2
Worldwide consumption of mercury used in ASGM practices.

Starting from the sixteenth century, amalgamation was the major technique used in the mining of gold (de Lacerda and Salomons, 1998; Hylander and Meili, 2003). Figure 2.2 depicts worldwide consumption of mercury used in practicing the ASGM. Telmer (2006) reported that approximately 15 million small-scale gold miners worldwide use mercury to trap the fine particles of this precious metal. The overall contribution of small-scale miners using the amalgamation can be understood by the fact that in 1995, China produced one-third of the total produced gold (105 t) from alluvial and coexisting ore by this technique. The negative side of this fact is that this mining releases tons of mercury into the environment every year, affecting millions of people's health (Lin et al., 1997). However, the inefficient recovery of mercury amalgam is inefficient for treating the remaining low-grade ores, and after the 1930s was replaced partially by the cyanidation process (UNEP, 2003). Its use by the small-scale ASGM operations may be due to the following reasons:

i. It is a rapid and easy method to extract gold from many alluvial ores under the existing field conditions. In the case study carried out by Telmer and Stapper (2007), the effective ore grade (what is recoverable by the miners) was about 0.1 g/t. The ore processed 100 t/day can produce a gravity concentrate of 10 kg of ore, with a concentration factor of 10,000 folds; containing 10 g gold would further require concentrating by 1000 folds. This can be done by manual gravity methods (like panning) but requires significant time and risks the loss of some gold (particularly the finer fraction). Recreational small-scale miners in Canada often spend more than 2 h panning up their concentrate; on the contrary, capturing the gold by amalgamation takes only 10 min and produces more certain results. Thus, the saved time in ASGM sites is invested to produce extra gold.

ii. Mercury is very independent, thereby eliminating the necessity of participating in undesirable and unfair labour practices.

iii. Mercury is highly effective to capture the gold under the practicing conditions found in ASGM sites. A centrifuge type of technique may be more effective than trapping gold with mercury, but at certain cost and infrastructure development.

iv. Mercury is relatively cheap and very accessible to small miners.

v. Small-scale miners are not always aware about the health risks posed by mercury and do not have scientific understanding of the alternatives.

vi. Mercury can give suitable results concentrating >10%–20% gold even when simple gravity methods cannot do it. A concentrate of 20% can directly be fed to smelter. Hence, it is used when capital (cash) is needed quickly (Agrawal, 2007).

Two methods primarily are used in the ASGM: (i) whole ore amalgamation and (ii) concentrate amalgamation. The amalgamation is often followed by a decomposition process to evaporate mercury, leaving gold to be recovered, and in some cases, cyanidation of amalgam tailings.

2.3.1 Whole Ore Amalgamation Process

Whole ore amalgamation is the process of bringing mercury into contact with the entire mining product, typically added either when the ore is being ground in mills or when the slurry produced from grinding is passed over a mercury coated copper plate. Amalgamating the whole ore uses mercury very inefficiently and consumes 3–50 folds mercury to the gold. Most of the mercury loss during this process initially occurs into the solid tailings which are often discharged directly into receiving waters and soils (Alpers and Hunerlach, 1999; Al et al., 2006; Shaw et al., 2006; Winch, 2006). Although little studied, the mercury in tailings subsequently leached with cyanide to enhance the gold recovery (a growing trend in some countries) also increases the discharge to the environment via the soluble cyano-mercuric complexes. Immediate emission to the atmosphere occurs when the amalgam is heated to separate the gold. In the simplest case, such as the use of mercury-coated copper plates, immediate losses to the atmosphere are therefore roughly equal to the amount of gold produced. However, there can be significant additional emissions to the atmosphere on a timescale of we–s to months from tailings and in particular from operations that employ cyanide. For example, 20 g of mercury consumed to produce 1 g of gold losses of 19 g mercury in tailing with immediate emission of 1 g mercury to the atmosphere (Sulaiman et al., 2007). Additional mercury is released to the atmosphere shortly thereafter by volatilization of CN-rich tailings.

2.3.2 Amalgamation of a Concentrate

Amalgamation of grinding concentrates can be done in batches, in which small gold particles are liberated to achieve a high yield of gold. Nevertheless, amalgamation of particles <700 mesh is difficult due to their suspension in the pulp. The optimum grinding practice must be determined, therefore, to release gold but not allow for the comminute of the released particles. Grinding sandy concentrates of gold with mercury in alkaline solution usually yields high recovery. Notably, the grinding time should be reduced if arsenic and other minerals that can consume mercury are present.

In cases where only a gravity concentrate is amalgamated, there are normally 1–2 unit mercury losses for each unit of gold produced, but significantly less if a mercury capturing system is used when the amalgam is burnt. For example, in Central Kalimantan, a ratio of 1.3:1.0 for Hg:Au is commonly used in tin amalgamation of a gravity concentrate produced by sluicing alluvial ore (Telmer and Stapper, 2007). In this case 0.3 g of mercury is discharged to water with the tailings and 1 g of mercury is emitted to the atmosphere by burning the amalgam, which is similar to what was recorded in Brazil (Sousa and Veiga, 2007). Sometimes the Zr-rich tailings are not discarded but processed further; amalgamated a second time to recover any residual gold, again consuming the mercury, followed by the emission to atmosphere during later industrial use.

2.3.3 Decomposition of Amalgams

The most common process to decompose amalgams is by increasing the temperature above 460°C when all mercury compounds are evaporated separating the gold (Angeloci, 2013). When the amalgams are burnt in open pans and mercury evaporated, one part of mercury is lost per one part of gold produced, as amalgams usually have 40%–50% Hg and 50%–60% other metals (Au, Ag, or Cu). Actually, since its application from the sixteenth century this is the most robust unit process of ASGM which directly threatens the environment by emitting it to the open atmosphere. The consequences of this are discussed separately.

2.3.4 Cyanidation of Amalgam Tailings

Tailings generated during amalgamation, which contain ~1 g/t gold with many folds of mercury therein, are treated to enhance the overall gold recovery, usually undergoing a cyanide leaching. By which, the gold along with the mercury and other metals forms various complexes with cyanide. Up to 90% gold from the tailings can be leached in the cyanide solution along with 40%–50% mercury, mainly due to a slower kinetics for mercury complexation with cyanide; hence, it keeps forming as the miners discharge the pulp into the receiving environment. Gold cyanide complexes formed during the leaching step are then adsorbed on activated carbon and removed from the leaching system without the need of filtration. Some species of $Hg(CN)_2$ are more easily adsorbed on the activated carbon than $[Hg(CN)_4]^{2-}$ (Adams, 1991). Furthermore, the gold cyanidation process itself occurs at pH levels between 10 and 11. Under this condition, it is expected that little mercury reports to the activated carbon. The tailings from which most of the gold was extracted with carbon are rich in mercury-cyanide complexes.

In Portovelo, Ecuador, where miners leach Hg-contaminated tailings (herein considered 100% of the mercury entering the cyanidation tanks), only 3.72% of the mercury is leached and removed by the activated carbon, and 51% of the mercury remains with the pulp of contaminated tailings in solution. The rest of the mercury is not dissolved and remains as droplets with the tailings to be discharged into the water streams. Gold is also extracted from the cyanide solution by precipitation with metallic zinc shavings (i.e. the Merrill-Crowe process). Precipitation with zinc is much more efficient at extracting mercury from cyanide solution than activated carbon is. Unfortunately, the portion of mercury precipitated along with the zinc is released to the atmosphere when miners irresponsibly evaporate the zinc shavings to obtain metallic gold.

2.4 Processing Centres of ASGM in Various Countries

The simple methodologies like centrifugal concentration that can achieve high gold recovery, even when compared with the environmentally destructive techniques of amalgamation, continue to be overlooked by artisanal miners. The ASGMs are generally more complicated, and processing at a central location may be economically beneficial to miners. With this, there is a greater likelihood for incorporation of these miners within a legal framework. Hence, several processing centres for ASGM have emerged worldwide in different locations, as shown in Figure 2.3.

Shamva (Zimbabwe). The concept of processing centres was successfully applied to the Shamva Mining Center in Zimbabwe, which was established

FIGURE 2.3
Worldwide processing centres of ASGM.

by the Intermediate Technology Development Group (ITDG) in 1989, providing access to a central mill for 200 small-scale miners (Mugova, 2001). Additionally, drilling and blasting services, ore transportation and technical support in areas of mine planning, method selection, safety and pollution control, information on legal issues, geology and metallurgy, and the selection and purchase of technical equipment is also provided by the Shamva. Miners pay approximately 6 USD/t and are individually responsible for their ore throughout the milling process. Usually gold concentrates up to 5 g/t and generated tailings occluded in particles or sulphide associated contain ~1 g/t of gold that can be processed by cyanidation. In another example, Proyecto Gama (Gestión Ambiental en la Minería Artesenal) has recently begun the design and engineering of a mercury-free artisanal processing plant in Peru. Gama intends to provide technical support for the project, but the cyanide plant will be constructed and operated by artisanal miners (Sousa et al., 2010; Cordy et al., 2011).

Nicaragua. Nicaragua has been experiencing a gold rush with official production of 7.46 t of gold in 2011, and it is believed a contribution of 1.24 t of gold was processed through the 20,000–30,000 artisanal miners, although not all of it by amalgamation. In Rozario, miners use 4″ dredges to pump the ore to sluice boxes with riffles concentrating only coarse gold. The processing centres charge the miners who are extracting hard rock ore 70 USD/day for use of one "rastra" with a capacity of 2 t/day. Rastras are grinding circuits, in which four rocks are dragged in circular movements by an electric motor over a bottom of rocks. As estimated, 160 rastras in operation can annually release 4–5 t mercury. Usually a copper plate (drip in 2 g/L NaCN solution) with mercury is kept at the discharge of the rastra to trap any gold leaving the grinding media. The Hg-contaminated tailing is sold for cyanidation leaching of gold by percolation in vat leaching. Analyses of one-ton tailing indicated 7.17 g Au and 10.3 g Hg in it, which leads to convert the mercury–cyanide complexes in the tailings reprocessing step. Analyses of total mercury in surrounding air (10 m away from the source) are found to be more than twice the exposure limit of $1000 ng Hg/m^3$ (WHO, 2008). Notably, it does not include the mercury when amalgams are burned without condensers (retorts) or filters.

Peru. Roughly estimated, approximately 80,000–150,000 artisanal gold miners are active in Peru, producing ~30 gold t/annum (Peru Support Group, 2008) and contributing an approximate emission of 20–40 t of mercury (Brooks et al., 2006). However, the ASGM sector in the country has continued to expand, particularly in Madre de Dios (Ashe, 2012). The processing centres located at the Piura region are extremely primitive and controlled by an association of artisanal miners. Mercury is introduced in "chanchas" (small ball mills) or, "quimbaletes" (which are large pieces of stone rocked back and forth, by one or more people to grind the ore with mercury) to amalgamate the whole ore without any prior (gravity) concentration. In this practice, 30%–40% mercury loss can be observed. In Piura, 80% of the miners are currently selling the ore to be processed in Nazca or Arequipa (South Peru). In many cases, the ores are transported to the processing centres in Portovelo (Ecuador) due to lack of knowledge of how to process the complex sulphide ores, lack of processing equipment, and lack of water and electricity. The business of moving Hg-contaminated tailings to process with cyanide is also transferring the risk of mercury mobility and bioaccumulation to other regions.

Colombia. Approximately 200,000 ASGM annually produce about 30 t gold in Colombia, resulting with the country being the largest mercury polluter per capita (~150 t Hg) exclusively from artisanal mining. The processing centres in Antioquia use small ball mills, like the "chanchas" (locally known as "cocos") to amalgamate whole ore (60–70 kg ore with 100–120 g mercury) without preconcentration. Tailings with grades up to 5 g/kg of Hg indicate that a substantial part of the mercury added to the "cocos" is pulverized and lost. Due to the historical presence of guerrillas in rural areas, the entire processing is carried out in urban areas and causes $40,000 ng/m^3$ Hg to be released into the surrounding air (Cordy et al., 2011). An open selling of amalgam with 50% Hg severely contributes to the urban pollution. Leaching of tailings in cyanide followed by precipitation with zinc and then evaporation in an open furnace to recover the gold again creates problems for the environment.

Indonesia. An unprecedented gold rush in the last 10 years resulted in 250,000 miners with ~800 artisanal mining sites in Indonesia, annually producing 40–60 t gold. It results in the release of 130–160 t mercury into the environment (Telmer and Veiga, 2008); however, recent estimates are between 280–560 t. Such a huge release is mainly due to the common practice of whole ore amalgamation in small ball mills, due to the presence of very fine gold that cannot be highly recovered by gravity concentration. It appears that similar practices of mining and processing are found all over Indonesia. A small group of miners commonly extract 100–1000 t/day of partially weathered saprolitic ore from shafts up to 20 m deep. The ore is delivered to a processing centre where it is ground into small ball mills (load of 20–40 kg of ore) and dosed within 300–500 g Hg and river cobbles or steel balls or rods. The grinding step lasts to 5 h, and the fine product is discharged into plastic bowls where the amalgam is separated by panning and ultimately burned with a torch in the open air. Only 30% gold can be recovered by this process (Veiga et al., 2009) and the material is very suitable for cyanidation, albeit the gold obtained by amalgamation is produced much faster than cyanidation (Krisnayanti et al., 2012).

Most miners in Indonesia sell or rent cyanidation tanks to process the Hg-contaminated tailings. The process consists of aeration for 2–3 days without agitation, after which activated charcoal is added to the leaching tanks. After three adsorption cycles, the charcoal can be recovered by screening. Most operators do not have knowledge about the activated charcoal elution process; hence, they burn it in drums, releasing contaminants (including Hg) to the atmosphere. The cyanidation tailings, along with mercury in solution, are deposited in rudimentary ponds or simply discharged into the rivers, causing aquatic problems (Castilhos et al., 2006).

Ecuador-Chile. It is estimated that the approximately 100,000 artisanal miners actively working in Ecuador (Sandoval, 2001) contribute ~75% of the total gold production of 25 t/year. In 2010, there were about 110 processing centres in the Portovelo-Zaruma region, which has been reduced to 87 centres. In the nearby town of Ponce-Enriquez, 200 km from Portovelo, it is estimated that an additional 48 processing centres opened in 2012 processing 3000 t of ore every day. About 10% of the processing centres process <10 t/day, 60% process 10–50 t/day, 25% process 50–100 t/day, and only 5% process >100 t/day, and they usually have their own captive mining operations. A large majority of the processing centres processes material in Chilean mills that typically consist of two or three heavy cement wheels with steel rims connected to a 20-hp electric motor. The wheels rotate over a 25-cm-wide, 2-inch-thick steel plate to crush and grind the material below 0.2 mm. The ground material is then concentrated in sluice boxes with wool carpets and the concentrate, around 200 kg, is then either panned to reduce the mass (15–20 kg) before being amalgamated manually in a pan, or directly leached with cyanide. In the first case, the amalgamation in a "batea" pan (known as "platon") contains a small amount of panned concentrate and represents low mercury losses with tailings (i.e. only 1.4% of the mercury introduced in the process). The amount of gold extracted by the sluice boxes fluctuates between 40% and 50% (Velasquez et al., 2010); however, after panning and amalgamation of the sluice concentrates, miners take home only 20%–30% of the gold from their ore.

All Hg-contaminated tailings are leached with cyanide and final tailings are discharged into the local rivers, estimated as ~650 kg Hg and 6000 t cyanide (Guimarães et al., 2011). The introduction of flotation at some processing centres to concentrate fine gold and copper sulphides has improved the process. On one hand, the Cu minerals can be sold; whereas on other side use of mercury and/or cyanide can be avoided since the smelters in Peru and China pay for the gold in concentrates. Despite the visible evolution of the centres, many still dispose mercury and cyanidation tailings into the local rivers and other receiving environments.

2.5 Environmental Impact by ASGM Activities and Associated Hazards

Mercury and cyanide are released to the environment during ASGM activities including whole ore amalgamation, gravity concentration, decomposition of amalgams, use of cyanide in mercury contaminated tailing, and dumping of mercury and cyanide in drainage (Veiga et al., 2006). During the amalgamation process, some mercury goes directly into water bodies as entrapped droplets or adsorbed onto sediment grains. Subsequently, the heating of amalgam (without fume hood or retort) emits mercury to the atmosphere. Naturally occurring mercury in soils and sediments is also eroded by sluicing and dredging, becoming remobilized and bioavailable in receiving waters (Telmer et al., 2006). When the amalgam tailing is treated by cyanidation, the mobilization of soluble cyano-mercuric complexes in water-bodies is possible, because mercury species can be easily methylated in the aquatic environment (Rodrigues-Filho et al., 2004; Castilhos et al., 2006; Sousa and Veiga, 2007; McDaniels et al., 2010; Guimarães et al., 2011).

The severe threat of mercury to the environmental release can be understood by an example of whole ore amalgamation operation (Suliman et al., 2007). If a consumption of 20 g mercury can produce 1 g gold, then 19 g of mercury is lost to the tailings and 1 g of mercury is immediately emitted to the atmosphere. However, additional mercury is released to the atmosphere shortly thereafter from (i) volatilization from cyanide rich tailings; (ii) cyanidation gold is adsorbed from the solution by activated carbon. Mercury is also unavoidably adsorbed. To recover the gold, the carbon is burnt and so any adsorbed mercury is emitted at that time; (iii) the "ash" produced by burning the activated carbon is often re-amalgamated with mercury; this amalgam is also thermally decomposed to produce the gold, releasing an additional amount of mercury to the atmosphere equal to the total gold produced. In such cases, immediate emissions to the atmosphere are minimally greater than the total gold produced, and this includes the amount of gold produced via cyanide leaching.

2.5.1 Associated Hazards of ASGM

The associated hazards of ASGM can be categorized as chemical, biological, physical, and psychosocial hazards.

2.5.1.1 Chemical Hazards

Miners are susceptible to inhaling, absorbing, and ingesting chemicals throughout the processing of ASGM; the most common chemical exposures are to mercury used in amalgamation and cyanide used for processing the amalgam tailings including chemical dust and gaseous hazards. Individuals working in or living near ASGM activities can be heavily exposed to mercury vapor and contaminated water that often exceeds the uptake limit fixed by the World Health Organization (United Nations Environment Program, 2012; Gibb and O'Leary, 2014). Due to its high volatility, elemental mercury can transform from its liquid state into vapours even at room temperatures (World Health Organization, 2003). Only small amounts of ingested elemental mercury that can easily be adsorbed in the gastrointestinal system can cause intoxication, which manifests in neurological, kidney, and autoimmune impairment (World Health Organization, 2013). Symptoms may intensify and/or become irreversible as exposure duration and concentration increase. Acute inhalation can directly affect the lung, causing irritation, chemical pneumonitis, pulmonary edema with consequent chest tightness, and respiratory distress (Dart and Sullivan, 2004; Agency for Toxic Substances and Disease Registry, 2014), which may lead to respiratory failure and death (Landrigan and Etzel, 2013). Under certain environmental conditions, mercury released into the environment can be transformed into an organic compound, namely methylmercury that enters into the food chain via the methylmercury containing fish, and a greater lipid solubility of methylmercury is easily absorbed into the bloodstream via the gastrointestinal tract. It can cause neurologic disease including tingling in the extremities, headaches, ataxia, dysarthria, visual field constriction, blindness, hearing impairment, psychiatric disturbance, muscle tremor, movement disorders, paralysis, and death (Gibb and O'Leary, 2014).

Besides the hazards generated directly by the chemicals, the small diameter of silica particles in the dusts generated during drilling, crushing, and blasting readily can be inhaled and deposited in the pulmonary system. This is toxic to lung tissue and to the immune system, causing progressive scarring and increased susceptibility to infectious agents, in particular tuberculosis (Rees and Murray, 2007; Guha et al., 2011; Gottesfeld et al., 2015). The presence of other minerals in the dust such as $FeAsS/PbS/CuFeS_2$ can also be hazardous (Molina-Villalba et al., 2015; Medecins Sans Frontieres, 2012). An incident of lead poisoning in Zamfara, Nigeria was a tragic reminder of such hazards. Over 400 children died from lead poisoning and more than 3000 people were poisoned.

Due to high gold recovery rate from cyanide at low cost, it is increasingly used in ASGM, but often after mercury has already been used, for example, on tailings (wastes). Mercury-cyanide compounds are easily dispersed in waters and, therefore, can enhance the mobility and/or bioavailability of mercury in the environment (United Nations Environment Programme, 2012). Cyanide interferes with human respiration at the cellular level and can cause severe and acute effects including rapid breathing, tremors, asphyxiation, and death (Lu, 2012). Chronic effects include neuropathological lesions, difficulty breathing, chest pain, nausea, headaches, and enlarged thyroid gland (Hinton et al., 2003b; Agency for Toxic Substances and Disease Registry, 2011). See Chapter 3 for a detailed discussion of the environmental aspects of cyanide.

2.5.1.2 Biological Hazards

Although ASGM communities are susceptible to a variety of infections, very common biological hazards affecting them are vector- and waterborne diseases and tuberculosis. Water and sanitation infrastructure is frequently lacking or inadequate in ASGM camps due to their remote location and transient activity. In some mining areas toilets are rare and pit latrines, if available, are usually shallow and can easily contaminate the water sources (Phillips et al., 2001), increasing the risk of waterborne diseases like cholera. Stagnant water provides a breeding environment for mosquitoes that carry diseases like malaria and dengue (de Santi et al., 2016). Water contamination associated with ASGM can occur in mines and hous›olds in the form of mine waste and chemical discharge.

2.5.1.3 Physical Hazards

The hazards due to robust working environment with heavy workloads, repetitive tasks, long working hours, and unsafe equipment cause musculoskeletal disorders, shoulder disorders, fatigue, and lower back pain (McPhee, 2004a,b). Physical hazards form a broad category that includes vibration, loud noise, heat and humidity, and radiation, all of which are present in ASGM. Miners experience shoulder disorders as a result of heavy lifting such as overhead work while suspending pipes and cables (Donoghue, 2004). They also experience chronic injury and fatigue from carrying heavy materials over long distances and bending over in awkward positions while panning or digging (Hinton et al., 2003b).

Physical trauma caused by overexertion results from uncomfortable postures and carrying out repetitive tasks using non-mechanized tools. Accidents cause burns, eye injuries, fractures, impalement, and, in some instances, physical dismemberment (Hinton, 2006; Navch et al., 2006; Long et al., 2015). In Ghana many cases of injuries arose primarily because of unsafe working conditions (Calys-Tagoe et al., 2015; Kyeremateng-Amoah and Clarke, 2015). The use of explosives can result in exposure to dangerous levels of dust, noise, and vibration and lead to asphyxiation and, in some cases, death due to acute traumatic injury (Harari and Harari Freire, 2013). Activities like mining, crushing, and milling are associated with elevated occupational and community noise levels, which often exceed WHO guideline limits for the prevention of hearing loss, which can result in hearing impairment, hypertension, ischemic heart disease, and stress (World Health Organization, 2011; Green et al., 2015). The labour-intensive ASGM can be compounded by extremely hot and humid working conditions resulting in heat stress that can cause dizziness, faintness, breathing difficulties, palpitations, and excessive thirst (Walle and Jennings, 2001).

2.5.1.4 Psychosocial Hazards

Social, cultural, and economic conditions can cause the emergence of psychosocial hazards. The migratory work nature of ASGM is believed to contribute to drug and alcohol abuse, which are seen as the ways to cope with difficulties (Hinton, 2003a,b,c; Donoghue, 2004; International Labor Organization, 2006; Thorsen, 2012). Drug and alcohol abuse can lead to violence against partners, co-workers, and community members and has been well documented in analogous scenarios (Hinton, 2006). However, some cases of violence are not caused by alcohol and can be a result of stressful working environment, and sometimes due to its illegal mining activities. Food security with nutritious diets is an important factor close to ASGM activities due to the deterioration in quality of agricultural lands (Hinton, 2006; Buxton, 2013). Changes in availability of disposable income among ASGM communities may also have an impact on quality of diets (Long et al., 2015).

2.6 Trends in Sustainable Development of ASGM

The sustainability of a process is dependent on several primary and secondary factors, in which the features of societal and socio-ecological factors are as important as the cost-effectiveness of the process (as pictorially presented in Figure 2.4). Notwithstanding the difficulties in defining a sustainable operation that exploits a non-renewable resource, the experiences from past and ongoing activities that can help to define the desirable conditions of ASGM include the following:

- Positive contribution of ASGM activities in development of rural areas and regional empowerment.
- Legal framework in harmony with the national mining sector policies.

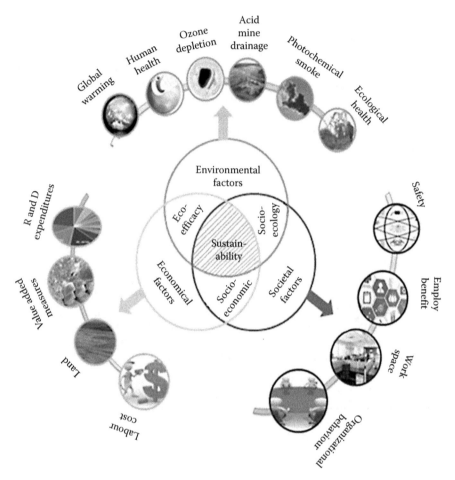

FIGURE 2.4
Interaction of factors affecting the sustainability of gold extraction process.

- Operation within international social standards, including the social security, occupational health and safety, and labour laws (that includes child labour), education and medical facilities, etc.
- Environmentally sound operation with scientific and mechanized inputs.
- Harmony between the small operations and large-scale mining operations.
- Ensuring high recovery yield, including a systematic development of the deposits and continuous operation.

Given the great importance to the rural workforce associated with ASGM, the potential of ASGM contribution to sustainable development is very high. Table 2.1 specifies the main factors needed to materialize on macro-, meso-, and microlevels for practicing a sustainable ASGM (MMSD, 2002).

2.6.1 Technical Aspects

Technical issues play a major role in ASGM and many of the problems can only be resolved by appropriate technical solutions. A prominent example is the mercury emissions by ASGM that can be solved by end-of-pipe technology (retorts, filters, and traps) or modifications in mineral processing. However, technical solutions can only be implemented by changing some of

TABLE 2.1

Factors Needed to Materialize on Macro-, Meso-, and Micro Levels for Sustainable ASGM

Level	Economic Aspects	Social Aspects	Environmental Aspects	Political Aspects
Macro level, State	Reception of taxes, royalties	Fair distribution of benefits	Minimization or elimination of water, ground, air and minerals resources	Need of liberal mining and economic policy Positive investment climate Participation of the mining sector in the planning of rural development
Meso level, community	Supply of services important for enterprises	Participation in mining activities Consideration of local interests during planning	Supply of services important for environment	Fluent dialogue between enterprise and government Existence of instruments and institutions for guidelines
Micro level, enterprise	Entrepreneurial competence and management knowledge Knowledge of the situation of the reserves Economic operation without support or subsidies through third parties	Quantified and motivated manpower Existence of an in-company programme for training and upgrading Social protection and safety for the miners	Controlled exploitation of the deposit Extraction of secondary products with environmental safety	Sound planning basis for utilization of the mineral, financial, material and human resources

the framework conditions, for which an interdisciplinary approach is crucial. Technical problems often require technical solutions but with an integral approach for their implementation. In contrast to the traditional definition of ASGM practicing low/non-mechanized activities, designing an alternative to simple stone mortar amalgamation mills is not high-cost rocket science. Conventional mining equipment is therefore frequently modified by the miners to fit their needs; unfortunately, in most cases suppressing security features (like water supply for drill hammers). The role of universities and research communities can be vital to it, as not much attention has been paid to ASTM due to its label as unorganized sector.

2.6.2 Policy and Legal Framework

In order to obtain a sustainable activity in ASGM that can potentially contribute to integrating the rural development with the formal economy, the governments need to develop a policy framework (MMSD, 2002) that can be based on the following strategies:

- Poverty alleviation
- Optimization of the business climate for the small mining sector
- Insurance of sustainability
- Stabilization of government revenues from the sector

There are a number of reasons for continuation of ASGM operation within the informal sector, mainly due to lacking knowledge of legal requirements. A lack of capacity on the part of governments to enforce penalties and to provide benefits, which should be associated with legalization, acts as a further disincentive to miners to be legalized (MMSD, 2002). Once the sector is acknowledged, governments need to develop a consistent and holistic sector policy. There are some examples in Peru, Colombia, Tanzania, and South Africa, where recent reforms in national policy have led to recognize the sector and attempt to enable a legal framework (MMSD, 2002). The main tasks in ASGM by the government bodies are seen in the issues related to the sustainable management and exploitation of mineral resources, promotion of investment into the sector, licensing the mining titles, and promoting their legalization.

Organization of the ASGM sector in the form of a community or society should be promoted to serve as a source of information and aid on all important issues related to the legal, fiscal, institutional, and administrative framework of the sector, the access to foreign markets, the activities within the sector (comprising not only mining itself but also the transformation, marketing, and exporting of the products). The local administrations should be encouraged to formalize the informal structures by coordination with an eye to a more effective and efficient, harmonized management of the local resources (Davidson, 1995; Keita, 2001).

Currently, in many countries the mining laws or other legal instruments do not support the development of small industries based on local mining production. This is especially valid for the production of informal ASGM, which is difficult to integrate into the formal economy. Nevertheless, prolongation of production lines should focus on creation of complementary activities through coordination with other groups. To eradicate the black markets for ASGM production, the government should include the informal production into the formal market.

Training resources for healthcare providers that directly address ASGM-related health issues are scarce. However, case studies, toxicology, and occupational health literature and publications from governmental and nongovernmental organizations do contain or suggest health components that could be developed further for use in this context (United National Industrial

TABLE 2.2

Cost and Benefits of ASGM Activities

Mining Costs	*Mining Benefits*
Exploitation of non-renewable resources	Possibility of exploiting smaller deposits
Losses, irrational working of high-grade material, incomplete exploitation, processing methods, transport	ASM achieves successful prospecting without high costs
	Working of abandoned pillars, tailings, etc.
Social Costs	*Social Benefits*
Precarious working conditions	Labour qualification
Negative health consequences (sickness, accidents)	Sources of income
Complicated dependency relations	Job creation
Insufficient social security and child labour	
Macro-Economic Costs	*Macro-Economic Benefits*
Conflicts; Due to land and water usage with governing bodies large-scale miners, indigenous population and landscape protection objectives	Mobilization of natural resources
Smuggling illegality (products and profit)	Tax collection and active effect for the balance of payments
No tax generation	Contribution to regional economic development by cash circulation, investment demand for products and services, mobility and structural consequences
Uncontrolled development due to lack of planned exploitation	Infrastructure development (road building, schools, energy supply) by small-scale mining and neighbouring population
	Comparative financial advantages (products with a high labour coefficient in countries with high labour availability)
	Relative stable product supply even with market fluctuations
	Contributes to product diversity and exports
	Substitutes imports

Development Organization, 2006). By using the principles of fair-trading, small-scale producers in developing countries are given the opportunity to trade their products under better selling terms and conditions. Financing of small mining projects should be tailor-made; non-conventional financing should be considered including self-financing, joint ventures and risk capital, and equity participation, as well as regular credit programs. A typical cost-benefit aspect of ASGM is shown in Table 2.2. An improved awareness of health hazards is needed to practice better, healthier, eco-friendly, and sustainable mining technologies (Hilson et al., 2007; Peplow and Augustine, 2007).

References

Adams, M.D. 1991. The mechanism of adsorption of $Hg(CN)_2$ and $HgCl_2$ on to activated carbon. *Hydrometallurgy*. 26: 201–210.

Agency for Toxic Substances and Disease Registry. 2014. Medical management guideline for mercury. http://www.atsdr.cdc.gov/mmg/mmg.asp?id=106&tid=24.

Agency for Toxic Substances and Disease Registry. 2011. Toxicological profile for boron. http://www.atsdr.cdc.gov/toxprofile/tp.asp?id=453&tid=80.

Agrawal, S. 2007. Community awareness on hazards of exposure to mercury and supply of equipment for mercury-cleaner gold processing technologies in Galangan, Central Kalimantan, Indonesia. Final report, UNIDO Project No. EG/GLO/01/G34, Contract no. 16001054/ML.

Al, T.A., Levbourne, M.I., Maprani, A.C., MacQuarrie, K.T., Dalziel, J.A., Fox, D., Yeats, P.A. 2006. Effects of acid sulfate weathering and cyanide-containing gold tailing on the transport and fate of mercury and other metals in Gossan Cre–: Murray Brook Mine, New Brunswick, Canada. *Applied Geochemistry*. 21: 1969–1985.

Alpers, C.N., Hunerlach, M.P. 2000. Mercury contamination from historic gold mining in California. USGS FS062-00, 6 p.

Angeloci, S.G. 2013. Myths and realities in artisanal gold mining mercury contamination. MASc thesis. Department of Mining Engineering. University of British Columbia, Vancouver, BC, p. 180.

Ashe, K. 2012. Elevated mercury concentrations in human of Madre de Dios, Peru. *PloS One*. 7(3): e33305.

Barry, M. 1996. Regularizing informal mining: A summary of the proceedings of the international roundtable on artisanal mining, IEN Occasional Paper no. 6.

Brooks, W.E., Sandoval, E., Yepez, M.A., Howard, H. 2006. Peru mercury inventory 2006. Open file report 2007-1252, U.S. Department of the Interior, U.S. Geological Survey, 55p.

Buxton, A. 2013. *Responding to the Challenge of Artisanal and Small-Scale Mining. How Can Knowledge Networks Help?* International Institute for Environment and Development, London.

Calys-Tagoe, B.N., Ovadje, L., Clarke, E., Basu, N., Robins, T. 2015. Injury profiles associated with artisanal and small-scale gold mining in Tarwwa, Ghana. *International Journal of Environmental Research and Public Health*. 12(7): 7922–7937.

Castilhos, Z.C., Rodrigues-Filho, S., Rodrigues, A.P., Villas-Bôas, R.C., Siegel, S., Veiga, M.M., Beinhoff, C. 2006. Mercury contamination in fish from gold mining areas in Indonesia and human health risk assessment. *Science of the Total Environment*. 368: 320–325.

Cordy, P., Veiga, M.M., Salih, I., Al-Saadi, S., Console, S., Garcia, O., Mesa, L.A., Velasquez-Lopez, P.C., Roeser, M. 2011. Mercury contamination from artisanal gold mining in Antioquia, Colombia: The world's highest per capita mercury pollution. *Science of the Total Environment*. 410: 154–160.

Dart, R.C., Sullivan, J.B. 2004. Mercury. In: Dart, R.C. et al. (Eds), *Medical Toxicology*, 3rd Ed. Lippincott Williams & Wilkins, Philadelphia, PA, pp. 1437–1448.

Davidson, J. 1995. Enabling conditions for the orderly development of artisanal mining with special reference to experiences in Latin America. In: *International Roundtable on Artisanal Mining*, World Bank, Washington, DC.

De Lacerda, L.D., Salomons, W. 1998. *Mercury from Gold and Silver Mining: A Chemical Time Bomb*. Springer, Heidelberg, Berlin, Germany.

de Santi, V.P., Dia, A., Adde, A., Hyvert, G., Galant, J., Mazevet, M., Nguyen, C., Vezenegho, S.B., Dusfour, I., Girod, R., Briolant, S. 2016. Malaria in French Guiana linked to illegal gold mining. *Emerging Infectious Diseases*. 22(2): 344–346.

Donoghue, A. 2004. Occupational health hazards in mining: An overview. *Occupational Medicine*. 54(5): 283–289.

Gibb, H., O'Leary, K.G. 2014. Mercury exposure and health impacts among individuals in the artisanal and small-scale gold mining community: A comprensive review. *Environmental Health Perspectives*. 122(7): 667–672.

Gottesfeld, P., Andrew, D., Dalhoff, J. 2015. Silica exposures in artisanal and small-scale gold mining in Tanzania and implications for tuberculosis prevention. *Occupational and Environmental Hygiene*. 12(9): 647–653.

Green, A., Jones, A.D., Sun, K., Nietzel, R.L. 2015. The association between noise, cortisol and heart rate in a small-scale gold mining community: A pilot study. *International Journal of Environmental Research and Public Health*. 12: 9952–9966.

Guha, N., Straif, K., Benbrahim-Tallaa, L. 2011. The IARC monographs on the carcinogenicity of crystalline silica. *La Medicina del Lavoro*. 102(4): 310–320.

Guimarães, J.R.D., Betancourt, O., Miranda, M.R., Barriga, R., Cueva, E., Betancourt, S. 2011. Long-range effect of cyanide on mercury methylation in a gold mining area in southern Ecuador. *Science of the Total Environment*. 409: 5026–5033.

Harari, R., Harari Freire, F. 2013. Safety and health in mining in Ecuador. In: Elgstrand, K., Vingard E., (Eds) *Occupational Safety and Health in Mining. Anthology on the Situation in 16 Mining Countries*. Occupational and Environmental Medicine at Sahlgrenska Academy, University of Gothenburg, Gothenburg, Sweden, pp. 171–178.

Hilson, G., Hilson, C.J., Pardie, S. 2007. Improving awareness of mercury pollution in small-scale gold mining communities: Challenges and ways forward in rural Ghana. *Environmental Research*. 103(2): 275–287.

Hinton, J. 2006. *Communities and Small-Scale Mining: An Integrated Review for Development Planning*. World Bank, Washington, DC.

Hinton, J., Veiga, M.M., Beinhoff, C. 2003a. Women and artisanal mining: Gender roles and the road ahead. In: Hilson, G. (Ed.), *The Socio-Economic Impacts of Artisanal and Small-Scale Mining in Developing Countries*. A.A. Balkema Publishers, Lisse, The Netherlands, pp. 161–203.

Hinton, J., Veiga, M.M., Beinhoff, C. 2003b. Women, mercury and artisanal gold mining: Risk communication and mitigation. *Journal de Physique IV*. 107: 617–620.

Hinton, J.J., Veiga, M.M., Veiga, A.T. 2003c. Clean artisanal gold mining: A utopian approach? *Journal of Cleaner Production*. 11: 99–115.

Hylander, L.D., Meili, M. 2003. 500 years of mercury production: Global annual inventory by region until 2000 and associated emissions. *The Science of the Total Environment*. 304: 13–27.

International Labour Organization. 2006. *Minors Out of Mining! Partnership for Global Action Against Child Labour in Small-Scale Mining*. International Labour Organization, Geneva, Switzerland.

Jennings, N. 1999. Social and labour issues in small-scale mines. In: *Report for Discussion at the Tripartite Meeting on Social and Labour Issues in Small-Scale Mines*, 17–21 May, Geneva.

Keita, S. 2001. The contribution of Sadiola gold mining project to poverty reduction and the development of local mining communities. Paper presented at MMSD (Mining, Minerals and Sustainable Development Project), London, p. 6.

Krisnayanti, B.D., Anderson, C.W.N., Utomo, W.H., Feng, X., Handayanto, E., Mudarisna, N., Ikram, H. 2012. Assessment of environmental mercury discharge at a four-year-old artisanal gold mining area on Lombok Island, Indonesia. *Journal of Environmental Monitoring*. 14: 2598–2607.

Kyeremateng-Amoah, E., Clarke, E. 2015. Injuries among artisanal and small-scale gold miners in Ghana. *International Journal of Environmental Research and Public Health*. 12: 10886–10896.

Landrigan, P.J., Etzel, R.A. 2013. *Textbook of Children's Environmental Health*. Oxford University Press, Oxford.

Lin, Y., Guo, M., Gan, W. 1997. Mercury pollution from small gold mines in China. *Water, Air and Soil Pollution*. 97: 233–239.

Long, R.N., Sun, K., Neitzel, R.L. 2015. Injury risk factors in a small-scale gold mining community in Ghana's Upper East Region. *International Journal of Environmental Research and Public Health*. 12(8): 8744–8761.

Lu, J.L. 2012. Occupational health and safety in small scale mining: Focus on women workers in the Philippines. *Journal of International Women's Studies*. 13(3): 103–113.

Malm, O. 1998. Gold mining as a source of mercury exposure in the Brazilian Amazon. *Environmental Research*. 77: 73–78.

McDaniels, J., Chouinard, R., Veiga, M.M. 2010. Appraising the global mercury project: An adaptive management approach to combating mercury pollution in small-scale gold mining. *International Journal of Environment Pollution*. 41: 242–258.

McPhee, B. 2004a. Ergonomics in mining. *Occupational Medicine*. 54: 297–303.

McPhee, M.E. 2004b. Generations in captivity increases b‹avioral variance: Considerations for captive breeding and reintroduction programs. *Biological Conservation*. 115(1): 71–77.

Medecins Sans Frontieres. 2012. Lead poisoning crisis in Zamfara State northern Nigeria. MSF briefing paper 2012. Medecins Sans Frontieres. http://www.doctorswithoutborders.org/sites/usa/files/Lead%20Poisoning%20Crisis%20in%20Zamfara%20State%20Northern%20Nigeria.pdf, Accessed 05 February 2016.

MMSD-Mining Minerals and Sustainable Development. 2002. *Breaking New Ground*. International Institute for Environment and Development and World Business Council for Sustainable Development, London, p. 441.

Molina-Villalba, I., Lacasaña, M., Rodríguez-Barranco, M., Hernández, A.F., Gonzalez-Alzaga, B., Aguilar-Garduño, C., Gil, F. 2015. Biomonitoring of arsenic, cadmium, lead, manganese and mercury in urine and hair of children living near mining and industrial areas. *Chemosphere*. 124: 83–91.

Mugova, A. 2001. The Shamva mining center project. Paper presented at MMSD (Mining, Minerals and Sustainable Development Project), London, p. 6.

Navch, T., Bolormaa, T., Enhtsetseg, B., Khurelmaa, D. 2006. *Informal Gold Mining in Mongolia—A Baseline Survey Report Covering Bornuur and Zaamar Soums, Tuv Aimag.* International Labour Organization, Geneva, Switzerland.

Peplow, D., Augustine, S. 2007. Community-directed risk assessment of mercury exposure from gold mining in Suriname. *Revista Panamericana de Salud Pública.* 22: 202–210.

Peru Support Group. 2008. The great water debate: Cause and effect in Peru. Update extra June 2008. Peru Support Group, London. http://www.perusupportgroup. org.UK/ resources.html.

Phillips, L.C., Semboja, H., Shukla, G.P., Swinga, R., Mutagwaba, W., Mchwmpaka, B. 2001. Tanzania's precious minerals boom: Issues in mining and marketing. Research paper. US Agency for International Development, Bureau for Africa, Office of Sustainable Development, Washington, DC.

Rees, D., Murray, J. 2007. Silica, silicosis and tuberculosis. *International Journal of Tuberculosis and Lung Disease.* 11(5): 474–484.

Rodrigues, F.S., Castilhos, Z.C., Santos, R.L.C., Yallouz, A.V., Nascimento, F.M.F., Egler, S.G. 2004. Environmental and health assessment in two small scale gold mining areas—Indonesia—Sulawesi and Kalimantan. CETEM's Technical final report to UNIDO-Reserved, p. 211.

Sandoval, F. 2001. Small scale mining in Ecuador, Report to MMSD—Environmental and Sustainable Development No. 75.

Shaw, S.A., Al, T.A., Macquarrie, K.T.B. 2006. Mercury mobility in unsaturated gold mine tailings, Murray Brook Mine, New Brunswick, Canada. *Applied Geochemistry.* 21: 1986–1998.

Sousa and Veiga. 2007. Brazil country report. UNDP/GEF/UNIDO Project EG/ GLO/01/G34. Final report.

Sousa, R.N., Veiga, M.M., Klein, B., Telmer, K., Gunson, A.J., Bernaudat, L. 2010. Strategies for reducing the environmental impact of reprocessing mercury contaminated tailings in the artisanal and small-scale gold mining sector: Insights from Tapajos River Basin, Brazil. *Journal of Cleaner Production.* 18: 1757–1766.

Suliman, R., Baker, R., Susulorini, B., Telmer, K., Spiegel, S. 2007. Indonesia country report. UNDP/GEF/UNIDO Project EG/GLO/01/G34. Final report.

Telmer, K. 2006. Mercury and small scale gold mining—Magnitude and challenges worldwide. Powerpoint presentation. In: *International Conference on Managing the International Supply and Demand of Mercury,* European Commission in Brussels, Belgium, October.

Telmer, K., Stapper, D. 2007. Evaluating and monitoring small scale gold mining and mercury use: Building a knowledge-base with satellite imagery and field work. UNDP/GEF/UNIDO Project EG/GLO/01/G34 Final Report.

Telmer, K., Stapper, D., Costa, M.P.F., Ribeiro, C., Veiga, M.M. 2006. Knowledge gaps in mercury pollution from gold mining. In: *Book of Abstracts of the 8th International Conference on Mercury as a Global Pollutant,* Madison, WI.

Telmer, K., Veiga, M.M. 2008. World emissions of mercury from small scale artisanal gold mining and the knowledge gaps about them. In: Pirrone. N., Mason, R. (Eds), *Mercury Fate and Transport in the Global Atmosphere: Measurements, Models and Policy Implications*. Springer, New York, pp. 96–129.

Thorsen, D. 2012. Children working in mines and quarries: Evidence from west and central Africa. UNICEF West and Central Africa Regional Office, Dakar-Yoff, Senegal.

UNEP. 2003. Global mercury assessment, Chapter: 7 Current production and use of mercury. United Nations Environmental Programme. United States Geological Survey, Reston, VA.

United Nations Environment Program. 2012. A practical guide: Reducing mercury use in artisanal and small-scale gold mining. United Nations Environment Programme, Geneva, Switzerland.

United Nations Environment Program. 2014. Minamata convention on mercury. http://www.mercuryconvention.org/Convention/tabid/3426/Default.aspx.

United Nations Environment Program. 2015. Developing a National Action Plan to reduce, and where feasible, eliminate mercury use in artisanal and small scale gold mining: Working draft. United Nations Environment Programme, Geneva, Switzerland.

United Nations Industrial Development Organization. 2006. Global mercury project: Environmental and health assessment report. United Nations Industrial Development Organization, Vienna, Austria.

Veiga, M.M., Nunes, D., Klein, B., Shandro, J.A., Velasquez-Lo, P.C., Sousa, R.N. 2009. Mill leaching: A viable substitute for mercury amalgamation in the artisanal gold mining sector? *Journal of Cleaner Production*. 17(15): 1373–1381.

Veiga, M.M., Maxson, P., Hylander, L. 2006. Origin of mercury in artisanal gold mining. *Journal of Cleaner Production*. 14: 436–447.

Velasquez-Lopez, P.C., Veiga, M.M., Hall, K. 2010. Mercury balance in amalgamation in artisanal and small-scale gold mining: Identifying strategies for reducing environmental pollution in Portovelo-Zaruma, Ecuador. *Journal of Cleaner Production*. 18: 226–232.

Walle, M., Jennings, N. 2001. Safety and health in small-scale surface mines: A handbook. International Labour Office (ILO) Working paper 168, ILO Publications, Geneva, Switzerland.

Winch, S., Parsons, M., Mills, H., Fortin, D., Lean, D., Kostka, J. 2006. Mercury speciation and sulfate-reducing bacteria in mine tailings. In: *Book of Abstracts of the 8th International Conference on Mercury as a Global Pollutant*, Madison, WI.

World Health Organization. 2003. Elemental mercury and inorganic mercury compounds: Human health aspects. Concise International Chemical Assessment Document 50. World Health Organization, Geneva, Switzerland.

World Health Organization. 2008. Training modules and instructions for health-care providers. http://www.who.int/c›/capacity/training_modules/en/.

World Health Organization. 2011. Burden of disease from environmental noise: Quantification of healthy years lost in Europe. WHO Regional Office for Europe, Bonn, Germany.

World Health Organization. 2013. Mercury exposure and health impacts among individuals in the artisanal and small-scale gold mining (ASGM) community. World Health Organization, Geneva, Switzerland.

3

Cyanidation of Gold-Bearing Ores

Sadia Ilyas*

3.1 Development History of Gold Cyanidation

Cyanidation of gold was a milestone in gold metallurgy, which can be considered a golden development in the processing of gold bearing ores. Cyanidation has been a tremendous help in gold extraction because the previously used technology of amalgamation and chlorination yielded only 55%–65% extraction efficiency from the ores containing finer particles. This is the reason that approximately one billion tons of gold bearing ores are currently processed by cyanidation (the largest tonnage of any mineral raw material for chemical treatment); hence, the history of this process is interesting to know.

The activities of chemists during the 18th century led to the discovery of cyanide; gold cyanidation was discovered in the late 18th and early 19th century (significant developments are summarized in Table 3.1). In the 18th century, chemists were occupied with a number of blue-coloured compounds obtained by heating dried blood with potash [K_2CO_3] that when subsequently treated with iron vitriol ($FeSO_4$) precipitated an intense blue pigment. In 1704 the German alchemist Johann Conrad Dippel (1673–1734) made its accidental discovery in Berlin, given the name as 'Berlin blue' that was later referred as the 'Prussian blue.' This new pigment immediately displaced the naturally occurring ultramarine blue pigment due to its being a cheaper pigment. This first artificially manufactured pigment opened a new field of chemistry, belonging the cyanogen compounds. In 1752 the French chemist Pierre Joseph Macquer (1718–1784) discovered the yellow crystals of potassium ferrocyanide by the mixing of alkali with Berlin blue under boiling conditions and separating the iron oxide. In 1782 the Swedish chemist Carl Wilhelm Scheele heated the blue pigment with dilute H_2SO_4 to obtain an inflammable gas that dissolves in water and has an acidic reaction with litmus paper, named Berlin Blue acid or simply, the blue acid (this was also used during World War II in gas chambers). He also discovered

* Mineral and Material Chemistry Lab, Department of Chemistry, University of Agriculture Faisalabad, Pakistan.

TABLE 3.1

Some Important Developments in Cyanidation with Time (by Year)

Year	Significant Developments
1704	Discovery of 'Berlin/Prussian blue" via heating dry blood with potash and subsequent treatment with iron vitriol
1752	Discovery of potassium ferrocyanide via boiling Prussian blue with alkali to separate the iron and concentrated solution yields yellow crystals of $K_3[Fe(CN)_6]$
1782	Discovery of blue acid via heating Prussian blue with dilute H_2SO_4; the dissolution of gold with cyanide was known to the Swedish chemist Carl Wilhelm Scheele
1811	Discovery of HCN via liquefying the blue acid gas
1822	Discovery of potassium ferricyanide via action of chlorine into a solution of potassium ferrocyanide
1836	In gold electroplating the bath necessary was introduced for recovery of gold from cyanide solution
1843	The existence of cyanogen compounds in coal gas was discovered
1844–1887	Various aspects of gold cyanidation reactions were studied, in 1846 the role of atmospheric oxygen was established
1887	MacArthur-Forrest process was established for gold extraction via suspending the crushed ore in a cyanide solution, up to 96% of pure gold was recovered
1896	Formation of H_2O_2 as an intermediate product during cyanidation was observed
1900	Merrill-Crowe process introduced to improve the MacArthur-Forrest process by precipitating gold with Zn-dust instead of using the Zn-shavings

the dissolution of gold into cyanide solution in 1783. In 1811, the French chemist Joseph Louis Gay-Lussac (1778–1850) liquefied the gas (boiling point 26°C) and determined its composition as hydrogen cyanide, HCN. Heating the Berlin Blue with HNO_3 solution yielded a red/violet compound known as 'prussiate' in reference to Prussia in Germany. In 1822, Leopold Gmelin (1788–1853) prepared potassium ferricyanide in the form of deep red prisms, by passing Cl_2 into a potassium ferrocyanide solution, which became a commercial product in 1825. A few years later, production of potassium cyanide was started. In 1835 the formation of HCN in blast furnaces became known, and in 1843 the existence of cyanogen compounds in coal gas was discovered (in the 1950s cyanides were also identified in different forms and at different locations throughout a petrochemical complex). In 1836, George Elkington (1801–1865) and Henry Elkington (1810–1852) used cyanide in solution to prepare the bath necessary for electroplating gold. During the years 1845–1890, many processes were investigated to develop the understanding of the chemistry of cyanides, cyanogens, thiocyanates, cyanates, cyanamides, and other related compounds. The dissolution reaction of gold was studied by noted chemists of that time (Elsner in Germany, 1846; Michael Faraday in England, 1857). In 1887, John Stewart MacArthur (1856–1920) in Glasgow applied the knowledge to gold ores and patented the cyanidation process. After that, a number of different research activities were initiated at many universities. In 1896 the German chemist Guido Bodländer (1855–1904) at the University of Breslau (in Poland) confirmed the necessity of oxygen in gold dissolution as previously claimed by Elsner and Faraday and discovered that hydrogen peroxide was formed as an intermediate product during the dissolution of gold.

However, the cyanidation process for gold dissolution was a mystery for a long time due to following reasons:

- It was difficult to understand why gold, the most noble metal that cannot be attacked by any strong acid except hot aqua regia, can be dissolved at room temperature in a dilute solution of 0.01%–0.1% of cyanide salts.
- A strong solution of NaCN was found to be no better than a dilute solution of 0.01%–0.1% of cyanide, whereas other metals dissolve faster in a more concentrated acid.
- The necessity of oxygen was doubted, but alone it has no action whatsoever on gold.

The queries were answered after a long time of ~60 years after the discovery of the cyanidation process, when the electrochemical b›aviour of the process was realized as similar to a galvanic cell. This has been demonstrated by embedding a small gold sphere in a KCN gel to which air was introduced from one direction, corroding gold at the surface far away from the point of airflow. The surface less exposed to oxygen acted as anode while the surface in direct contact with oxygen acted as cathode. By this means, oxygen picks electrons from the gold surface while gold ions enter into the solution and are rapidly complexed by the cyanide ions therein.

With the analogy of the 'transmutation' of iron into copper, MacArthur used zinc shavings to precipitate gold from the cyanide solution. The process became more efficient when zinc dust was introduced by Charles W. Merrill (1869–1956) in around 1900. Further, the process was improved by Thomas B. Crowe who removed air from the solution by passing it through a vacuum tank before introducing the zinc; this became known as the Merrill–Crowe process (Habashi, 1967).

3.2 Cyanide Leaching of Gold

The gold cyanidation process for its extraction from ore bodies involves heterogeneous reactions at the interfaces of solid and liquid (Figure 3.1). The dissolution is supposed to follow these sequential steps:

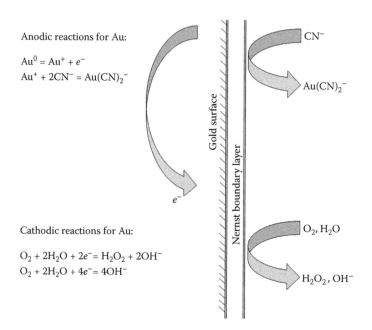

Anodic reactions for Au:

$Au^0 = Au^+ + e^-$
$Au^+ + 2CN^- = Au(CN)_2^-$

Cathodic reactions for Au:

$O_2 + 2H_2O + 2e^- = H_2O_2 + 2OH^-$
$O_2 + 2H_2O + 4e^- = 4OH^-$

FIGURE 3.1
A typical electrochemical leaching mechanism of gold in cyanide solution. (Modified from Habashi, 1967).

i. Oxygen solubilization via absorption in the cyanide solution
ii. Transportation of dissolved oxygen and cyanide to the solid-liquid interface
iii. Diffusion of reactants CN^- and O_2 onto solid surfaces through the Nernst boundary layer
iv. Occurrence of electrochemical reactions
v. Desorption of the solubilized gold-cyanide complex and other reaction products from the solid surface
vi. Transportation of the reaction products to the bulk solution

Electrochemically, the oxygen takes up electrons at one part of the gold surface, the cathodic zone; whereas, gold donates electrons at another part, the anodic zone. Accordingly, the dissolution rate is a function of cyanide concentration at low cyanide concentrations. At high cyanide concentrations, however, the dissolution rate depends on the soluble concentration of oxygen. Hence, the dissolution rate increases linearly with respect to increasing the concentration of cyanide until a maximum dissolution of gold is reached; beyond that, a slight retarding effect occurs. The activation energy for dissolving gold in a cyanide solution via mass-transport control is calculated in the range of 8–20 kJ/mol (Habashi, 1967).

TABLE 3.2

Stability Constant Values of Gold Complexes
(Marsden and House, 1992)

Gold Complexes	Stability Constant Value
$Au(CN)_2^-$	2×10^{38}
$Au(S_2O_3)_2^{3-}$	5×10^{28}
$Au[CS(NH_2)_2]^+$	2×10^{23}
AuI_2^-	4×10^{19}
$Au(SCN)_2^-$	1.3×10^{17}
$AuBr_2^-$	1×10^{12}
$AuCl_2^-$	1×10^9

The selectivity of free cyanide for gold dissolution along with extremely high stability of the cyanide complexes (illustrated in Table 3.2) are the main advantages of the process.

3.2.1 Various Theories of Gold-Cyanidation

Cyanidation has been a great help in gold metallurgy, and as such has been widely investigated. Several theories b›ind gold leaching in cyanide solution have been presented (Cornejo and Spottiswood, 1984; Habashi, 1987).

Elsner's oxygen theory: Elsner (1846) was the first to recognize that oxygen was essential for the dissolution of gold in cyanide solution:

$$4Au + 4NaCN + O_2 + H_2O = 4NaAu(CN)_2 + 4NaOH \qquad (3.1)$$

Janin's hydrogen theory: Janin (1888, 1892) disagreed with the necessity of oxygen for gold dissolution in cyanide solution, and described the hydrogen evolution as:

$$2Au + 4NaCN + 2H_2O = 2NaAu(CN)_2 + 2NaOH + H_2 \qquad (3.2)$$

Later, Maclaurin (1893) and Christy (1896) concluded the necessity of oxygen in cyanidation, and favoured Elsner's theory.

Bodländer's hydrogen peroxide theory: Bodländer (1896) suggested a two-step dissolution reaction for gold cyanidation, as shown in the following equations:

$$2Au + 4NaCN + O_2 + 2H_2O = 2NaAu(CN)_2 + 2NaOH + H_2O_2 \qquad (3.3)$$

$$2Au + 4NaCN + H_2O_2 = 2NaAu(CN)_2 + 2NaOH \qquad (3.4)$$

The formation of H_2O_2 as an intermediate product was supported by its detection during the cyanidation process. The overall equation of 3.3 and 3.4 is known as the Elsner's equation.

Cyanogen formation: Christy (1896) suggested the liberation of cyanogen gas formed by the action of oxygen with cyanide solution, which can be the active agent for attacking the gold. The reactions in two steps can be understood as:

$$\frac{1}{2}O_2 + 2NaCN + H_2O = (CN)_2 + 2NaOH \tag{3.5}$$

$$2Au + 2NaCN + (CN)_2 = 2NaAu(CN)_2 \tag{3.6}$$

Later, Skey (1897) and Park (1898) gave conclusive evidence that the aqueous cyanogen does not exert the least solvent action on gold.

Cyanate formation: MacArthur (1905) argued for the effectiveness of potassium cyanate in gold dissolution that is formed by the oxidation of cyanide with supplied oxygen. Nevertheless, the assumption was refuted by Green (1913) by showing that cyanate had no action on gold. Thermodynamic evidence given by Barsky et al. (1934) determined the free energies for the complexes of auro- and argento-cyanide ions, based on which the calculated free energy changes were in favour of Eisner and Bodländer's equations; whereas Janin's equation was thermodynamically not feasible.

Corrosion theory: The fact that the complexation of gold in cyanide solution is similar to the metal corrosion process was first revealed by Boonstra (1943). It was observed that the dissolved oxygen in cyanide solution is reduced to the hydrogen peroxide and hydroxyl ion, making the reaction of electrochemical nature. On that basis, Bodländer's equation was further divided into the following steps:

$$O_2 + 2H_2O + 2e^- = H_2O_2 + 2OH \tag{3.7}$$

$$H_2O_2 + 2e^- = 2OH^- \tag{3.8}$$

$$Au = Au^+ + e^- \tag{3.9}$$

$$Au^+ + CN^- = AuCN \tag{3.10}$$

$$AuCN + CN^- = Au(CN)_2^- \tag{3.11}$$

3.2.2 Factors Affecting the Cyanidation Process

The parametric effects on the gold cyanidation process have been studied widely and deliberated to get an enhanced efficacy.

3.2.2.1 Effect of Cyanide Concentration

The rate of gold leaching in cyanide solution increases linearly with respective increase in cyanide concentration until attaining the maximum point (Marsden and House, 1992; Kondos et al., 1995; Ling et al., 1996; Wadsworth et al., 2000; Deschênes et al., 2003). A further increase in cyanide concentration does not enhance the gold leaching; on the contrary, it shows a retarding effect and decreases the gold content in leach liquor as shown in Figure 3.2 (Maclaurin, 1893; Barsky et al., 1934). At a high concentration the cyanide ion undergoes a hydrolysis reaction (as shown Equation 3.12), resulting in an increase in alkalinity, which suppresses the leaching of gold.

$$CN^- + H_2O = HCN + OH^- \qquad (3.12)$$

However, Ling et al. (1996) showed that the retarding effect could be overcome by a change of concentration of cyanide at higher levels. The effect can be seen in Figure 3.3, in which the use of 400–500 ppm sodium cyanide results in a similar amount of gold leaching (Deschênes et al., 2003). A similar observation made by Ellis and Senanayake (2004) stated that the leaching rate could be enhanced with an increase in cyanide concentration, but becomes independent to the cyanide concentration if it exceeds 0.075% of potassium cyanide or 0.06% sodium cyanide in the lixiviant solution. In actuality, an excess of cyanide causes needless consumption of cyanide, which does not favour the leaching reactions (Kondos et al., 1995). The large amount of cyanide would create cyanide complexes with impurities. Conversely, a higher cyanide concentration possibly could be needed due to its competition with other minerals that may be present in the ore (Marsden and House, 2006). Moreover, a decrease in cyanide concentration limits its consumption, which ultimately controls the extra cost of effluent treatment at industrial scale

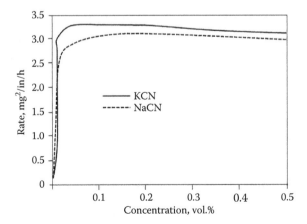

FIGURE 3.2
Effect of cyanide (KCN and NaCN) concentration on leaching rate of gold.

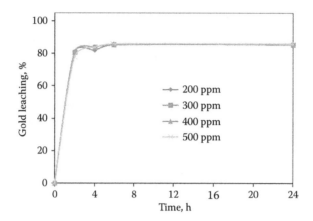

FIGURE 3.3
The leaching efficiency of gold as a function of time with respect to the varied concentration of NaCN (Deschênes et al., 2003).

(Ling et al., 1996; Deschênes et al., 2003). Therefore, a concentration of cyanide lesser or greater than the optimal level may negatively affect the gold cyanidation process.

3.2.2.2 Effects of the Eh-pH

The aqueous chemistry of gold in cyanide solution presented in Figure 3.4 shows the aurous-cyanide ions, $Au(CN)_2^-$, predominantly exist in wider pH ranges while the auric ions, Au^{3+}, exist in a very limited area. The stability field of Au^0 at relatively low potential value (Eh) covers the whole pH range, as does the stability of water. At high Eh values, gold can form insoluble oxides (hydrated auric oxide or gold peroxide) that are thermodynamically unstable and hence powerful oxidants. The oxidizing power of these compounds depends on the acidity of the system and declines with raise in pH. At very low Eh values, hydrogen cyanide (HCN) and cyanide ions (CN$^-$) are the stable species, the latter being predominant at pH > 9.24, whereas, the cyanate (CNO$^-$) is the only stable species at higher values of Eh.

As can be seen in Figure 3.4, the aurous cyanide complex [Au(CN)$_2$]$^-$ has a substantial stability, extending into large area of the Au-H$_2$O stability fields. The presence of extensive stability fields for this compound, especially at pH > 9.2, where formation of HCN can be totally avoided, makes the leaching feasible in cyanide solution. Nevertheless, the solid aurous cyanide, AuCN, and the auric cyanide complex [Au(CN)$_4$]$^-$ have been reported; the aurous cyanide complex [Au(CN)$_2$]$^-$ is the only stable complex of gold cyanidation. It has been found that the introduction of cyanide in aqueous systems drastically reduces the stability fields of zero valence gold and its oxides.

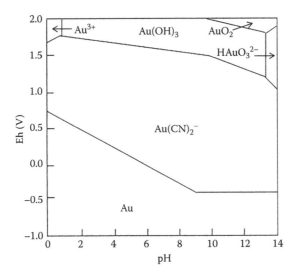

FIGURE 3.4
The Eh-pH diagram of the Au-CN-H_2O system at 25°C and 1 atm.

3.2.2.3 Effect of Alkali Addition

A purposeful addition of alkalis (i) prevents the hydrolysis loss of cyanide, (ii) prevents the cyanide loss by the action of carbon dioxide in air, (iii) decomposes bicarbonates in the mill water prior to the cyanidation, (iv) neutralizes the acidic compounds (such as ferrous salts, ferric salts, and magnesium sulphate) in the mill water before adding it to the cyanide circuit, and (v) neutralizes acidic constituents-pyrite, alkalis (CaO/NaOH/Na_2CO_3) which are added in the gold-cyanidation process. The use of lime additionally promotes the settling of fine particles to obtain a clear leach liquor that can be separated easily from the pulp.

Although the use of an alkali is essential in cyanidation, several researchers have revealed the addition of alkali like sodium hydroxide, and particularly calcium hydroxide, retards the cyanidation leaching of gold. Barsky et al. (1934) investigated the effects of calcium hydroxide and sodium hydroxide on the rate of gold dissolution in cyanide solutions containing 0.1% NaCN and found a decreased leaching rate when calcium hydroxide was added with a solution pH ~11. Leaching was almost nil at pH ≥ 12.2. The effect of NaOH was less pronounced when the leaching rate started was slower at pH > 12.5. However, leaching was more rapid at pH 13.4 using NaOH than in a solution containing similar concentration of cyanide by using calcium hydroxide at pH 12.2. Hence, the effect of calcium ion on leaching was then investigated by adding $CaCl_2$ and $CaSO_4$ to a cyanide solution. Neither of these salts affected the leaching rate to any appreciable extent. The solubility of oxygen in cyanide solutions with various amounts of Ca(OH)$_2$also did not show any appreciable difference.

Thus, it is concluded that a decrease in leaching rate using NaCN solutions caused by the addition of $Ca(OH)_2$ is not due to either lower solubility of oxygen or to the presence of calcium ions. Habashi (1967) attributed the retarding effect of $Ca(OH)_2$ by formatting calcium peroxide onto gold surfaces, which prevents the cyanidation reaction. Notably, calcium peroxide is supposed to form by the reaction of lime with H_2O_2 accumulating in the solution.

Moreover, cyanide exits as HCN gas in less alkali region (pH < 9.2) where the formation of insoluble AuCN along with hydrogen peroxide is possible, as shown in the reaction below:

$$2Au + 2HCN + O_2 = 2AuCN + H_2O_2 \tag{3.13}$$

To avoid the formation of AuCN, therefore, the cyanide solution should be alkaline which can control the decomposition of cyanide ions via hydrolysis (Eq. 3.12) and in presence of atmospheric CO_2 (Equation 3.14) as well.

$$CN^- + H_2CO_3 = HCN + HCO_3^- \tag{3.14}$$

The HCN forms insoluble AuCN; whereas, the adverse effect on leaching rate at very high pH is also avoidable. Thus, it is important to carefully optimize the leaching pH, which is usually between 11 and 12.

3.2.2.4 Effect of Oxygen Concentration

The Elsner's equation (Equation 3.1) shows the criticality of oxygen in gold cyanidation. The maximum dissolved oxygen content of a dilute cyanide solution is 8.2 ppm at 25°C. But cyanidation is performed at pH ~11 where the dissolved O_2 concentration remains ~6 ppm. A decrease in O_2 concentration below 4 ppm greatly reduces the leaching rate. In contrast, leaching increases remarkably when concentration of dissolved oxygen rises above 10 ppm. An oxygen-enriched operation (12–18 ppm O_2) for processing the gold-bearing sulphide ores is beneficial for achieving high throughput at a commercial level. The leaching of sulphide ores consumes oxygen, as shown in Equations 3.15–3.18; it is imperative to avoid a decrease in leaching rate associated with low content of dissolved oxygen in the system.

$$2MS + 2(x-1)CN^- + O_2 + H_2O = 2M(CN)_x^{(2-x)-} + 2CNS^- + 4OH^- \tag{3.15}$$

$$MS + 2OH^- = M(OH)_2 + S^{2-} \tag{3.16}$$

$$2M(OH)_2 + \frac{1}{2}O_2 + H_2O = 2M(OH)_3 \tag{3.17}$$

$$2S^{2-} + 2O_2 + H_2O = S_2O_3^{2-} + 2OH^- \tag{3.18}$$

Notably, pure oxygen was first introduced in the gold cyanidation process by Air Products, South Africa in the 1980s (Stephens, 1988) and then practised in Canadian plants as well (McMullen and Thompson, 1989). The Lac Minerals plants were the first to demonstrate the faster leaching rate associated with the dissolved oxygen and lead nitrate. Deschênes et al. (2003) investigated the effect of dissolved oxygen (as presented in Figure 3.5) where it was concluded that dissolved oxygen concentration is not much related to cyanide consumption; rather, a rapid leaching kinetics can be achieved.

The use of oxygen-assisted leaching was adopted quickly in industrial practice with the technological advancement for improving oxygen mass transfer (e.g., the use of Degussa's peroxide-assisted leach or pressure acid leaching, PAL and Kamyr's carbon-in-leach-with oxygen process, CILO) (Elmore et al., 1988; Loroesch et al., 1988; Revy et al., 1991; Kondos et al., 1995; Liu and Yen, 1995). Numerous efforts have been made to develop an efficient device for enhancing the oxygen dispersion (Jara and Harris, 1994; Sceresini, 1997; McLaughlin et al., 1999). The practice of continuous oxygen monitoring and control stabilizes process performance and compensates for disturbances related to changes in oxygen requirements. A better design of oxygen probes has added further robustness to the control and operation strategy in leaching (McMullen and Thompson, 1989). The FILBLAST Gas-Shear Reactor has been employed to improve gas mass transfer efficiency via an improved dissolved oxygen concentration and reduced oxygen consumption, which could enhance the leaching rate of gold in cyanide solution.

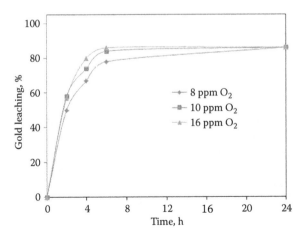

FIGURE 3.5
Effect of dissolved oxygen concentration on cyanide leaching of gold carried out at pH 11.2 with 500 ppm NaCN concentration and time 24 h. (Modified from Deschênes et al., 2003).

3.2.2.5 Effect of Particle Size

Particle size is a vital factor that affects gold cyanidation. A gold bearing ore requires fine grinding to liberate the encapsulated particles of precious metal from the ores and make them amenable to be leached in cyanide solution. Ling et al. (1996) reported that smaller particle size could improve the rate of gold dissolution due to larger surface area and longer contact time between solid and lixiviant. At the optimal condition of aeration and agitation, the maximum rate of gold leaching is determined to be $3.25\,mg/cm^2/h$. This equals a penetration of 1.68 microns on each side of a flat $1\,cm^2$ gold particle, or a total reduction in thickness of 3.36 microns hourly. At such a rate, a gold particle of 37 microns thickness would take about 11 h to completely leach out in the solution. In general, coarse free gold particles are removed by gravity concentration prior to cyanidation, as these particles might not completely dissolve in an acceptable duration for leaching. Fine grinding of ores, however, raises the cyanide consumption up a significant amount (Kondos et al., 1995). Similar observations have also been reported by Ellis and Senanayake (2004) and indicated that the leaching may be reduced with a decrease in particle size, as the rate of competing reactions and lixiviant using side reactions can be greater. Furthermore, grinding is a costlier operation; the finer and finer grinding can increase the cost several folds.

3.2.2.6 Effect of Temperature

In general, the reaction kinetics increases with respect to elevating the temperature. The temperature affects the gold cyanidation process in two ways:

  i. An increase in temperature increases the activity of cyanide solution, thus increasing the rate of gold leaching, and
  ii. The amount of dissolved oxygen decreases under ambient pressure, which decreases with increasing temperature.

Gold leached in cyanide solution (0.25% KCN) as a function of temperature (shown in Figure 3.6) showed the maximum efficiency at 85°C, albeit a half amount of dissolved oxygen at this temperature than the oxygen solubility at 25°C. A further elevation in temperature could result in slightly declined leaching, although the solution contains no dissolved oxygen at 100°C. Such a result is attributed to a lower capacity of an electrode to adsorb/retain hydrogen in a heated solution. Hence, the maximum opposing electromotive force (EMF) due to polarization becomes lesser for the heated solution until the EMF of gold leaching overbalances polarization and proceeds gold leaching even without dissolved oxygen in the cyanide solution. Polarization can be prevented either by oxygen; oxidizing the hydrogen at gold surface to favour the leaching at low temperatures; or by heat, dislodging the hydrogen from

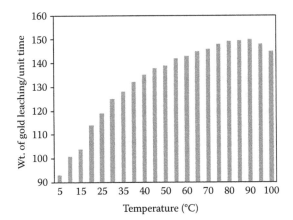

FIGURE 3.6
Effect of temperature on gold leaching rate with 0.25% KCN solution under aeriation. (Modified from Julian and Smart, 1921).

gold surface to favour leaching without the oxygen. The activation energy of gold leaching ranges from 2 to 5 kcal/mole, which is typically the diffusion-controlled reaction.

3.2.2.7 Effect of Agitation Speed

A proper contact between the solid mass and liquid lixiviant is a key factor in cyanidation leaching of gold. The rate of leaching depends on the thickness of the diffusive layer and the mixing pattern (Marsden and House, 1992). Consequently, the agitation must be of adequate speed to suspend all the ore particles in lixiviant solution. By increasing the stirring speed, the leaching kinetics increases. An intense mixing pattern reduces the thickness of the diffusive layer around ore particles, significantly improves the rate of mass transfer of oxygen and cyanide, and allows feasible saturation of pulp (Ellis and Senanayake, 2004).

3.2.2.8 Effect of Foreign Metal Impurities

Gold mostly occurs in native form, along with varying amounts of coexisting silver. Besides silver, several minerals characteristically coexist with gold, such as pyrite, galena, zinc blend, arseno-pyrite, stibnite, pyrrhotite, and chalcopyrite. While investigating the solubility of sulphide minerals, Hedley and Tabachnick (1958) reported the decreasing order of solubility as: pyrrhotite > marcasite > pyrite. The solubility of these minerals is pH dependent, dissolving orpiment, stibnite, and realgar at pH ~12 while orpiment dissolves appreciably at pH 10. Recent studies on stibnite, arsenopyrite, and realgar (Deschênes et al., 2000; Guo et al., 2004) obtained different findings that arsenopyrite does not react to a noticeable extent in cyanide solution.

At pH 11.5, realgar significantly interferes with gold leaching, reducing the kinetics by 40% in 4 h leaching. Various selenium minerals, magnetite, uraninite, and sometimes carbonaceous matter remain associated with the gold bearing ores. The most common gangue minerals (quartz, feldspar, micas, garnet, and calcite) are insoluble in cyanide solution, whereas some metallic minerals are soluble. The presence of carbonaceous matter is a matter of concern, as it significantly adsorbs the gold-cyanide complex and causes operational loss. The minerals that dissolve in cyanide solution influence the leaching by either accelerating or retarding effect.

3.2.2.8.1 Accelerating Effect

Pyrite and pyrrhotite are a flourishing variety of sulphur, and iron cyanide complexes into the leach liquor, including chalcopyrite forms of different cyanide complexes of copper (Aghamirian and Yen, 2005). The sulphide ions negatively affect the leaching rate of gold and cause surface passivation by diffusion layer (Marsden and House, 1992; Liu and Yen, 1995; Aghamirian and Yen, 2005; Dai and Jeffrey, 2006). The effect of sulphide minerals on gold leaching is shown in Figure 3.7. Liu and Yen (1995) have reported accelerated leaching kinetics when oxygen concentration is increased in the presence of pyrite and chalcopyrite.

The presence of small amounts of lead, mercury, bismuth, and thallium salts also accelerates the leaching. The electrode potential value of these metals in cyanide solution indicates that gold can displace these metal ions. The accelerating effect in the presence of these metal ions can be

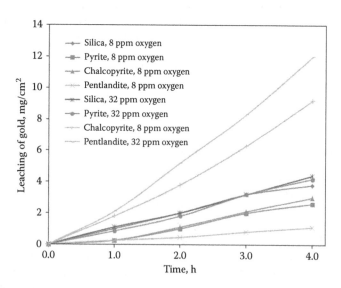

FIGURE 3.7
Effect of pyrite, pentlandite, chalcopyrite and oxygen on cyanide leaching of gold as a function of time.

corroborated to the alteration in the surface character of gold by alloying with displaced ions. A change in surface character may lead to a decreasing thickness of the boundary layer, through which the reactants diffuse to reach the metallic surface. The influence of lead sulphide on leaching rate differs somewhat from the sulphide-bearing minerals described earlier. An accumulation of PbS causes the improved leaching rate of gold under the optimal conditions. Arsenopyrite also has a positive effect on gold cyanidation (Liu and Yen, 1995), as shown in Figure 3.8. The following reactions indicate the acceleration of gold leaching by formation of various reaction products (Jin et al., 1998).

$$2Pb(OH)_2 + 5Au + 8CN^- = AuPb_2 + 4Au(CN)_2^- + 4OH^- \qquad (3.19)$$

$$2AuPb_2 + 2CN^- + 4OH^- = Au(CN)_2^- + 2PbO + 2H_2O + 5e^- \qquad (3.20)$$

$$Pb(OH)_2 + 4Au + 4CN^- = Au_2Pb + 2Au(CN)_2^- + 2OH^- \qquad (3.21)$$

$$3Pb(OH)_2 + 7Au + 12CN^- = AuPb_3 + 6Au(CN)_2^- + 6OH^- \qquad (3.22)$$

3.2.2.8.2 Retarding Effect

The retarding effect on the gold cyanidation process may occur due to one or more of the following reasons.

Consumption of oxygen from solution: Oxygen is necessary for gold cyanidation leaching. Any side reaction may cause it to be deprived of its

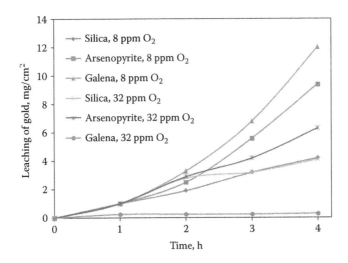

FIGURE 3.8
Effect of galena, arsenopyrite and oxygen on cyanide leaching of gold as a function of time. (Modified from Liu and Yen, 1995).

oxygen content in cyanide solution and will lead to a decrease in leaching rate. Pyrrhotite accompanying gold in its ores decomposes in an alkaline medium, yielding a series of reaction products which leads to oxygen depletion therein the solution.

$$FeS + 2OH^- = Fe(OH)_2 + S^{2-} \tag{3.23}$$

$$2Fe(OH)_2 + \frac{1}{2}O_2 + H_2O = 2Fe(OH)_3 \tag{3.24}$$

$$2S^{2-} + 2O_2 + H_2O = S_2O_3^{2-} + 2OH^- \tag{3.25}$$

Consumption of free cyanide from solution: Metal impurities like silver, copper, zinc, and iron associated with gold ore may dissolve in cyanide solution, causing depletion of cyanide content from the lixiviant. For example, a reaction with zinc can be written as:

$$ZnS + 4CN^- = Zn(CN)_4^{2-} + S^{2-} \tag{3.26}$$

Formation of thiocyanate in the presence of oxygen significantly loses cyanide contents as thiocyanate has no action on gold.

$$S^{2-} + CN^- + \frac{1}{2}O_2 + H_2O = CNS^- + 2OH^- \tag{3.27}$$

The auriferous ores containing quartz or aluminosilicates form colloidal silica and alumina in alkaline pH, along with precipitating the iron. These reaction products have strong adsorptive capacity for sodium cyanide, and thus retard the gold leaching. Moreover, the amount of cyanide required to complex with gold in an average ore is negligible but is significantly consumed by the sulphide of silver (as shown in Equation 28) usually present with gold minerals.

$$Ag_2S + 5NaCN + \frac{1}{2}O_2 + H_2O = 2NaAg(CN)_2 + NaCNS + 2NaOH \tag{3.28}$$

In the presence of CO_2 and insufficient free alkali, cyanide decomposes as:

$$2NaCN + CO_2 + H_2O = 2HCN + Na_2CO_3 \tag{3.29}$$

A large amount of lead ions causes a retarding effect by the formation of an insoluble film of $Pb(CN)_2$ onto the gold surface.

3.2.2.9 Effect of Film Formation onto Gold Surface

Sulphides: The retarding effect generated in the presence of a sulphide ion (shown in Figure 3.9a) indicates that a sulphide ion as small as 0.5 ppm can retard the leaching. This cannot be accounted for the depletion of cyanide or oxygen content; however, the formation of an insoluble aurous sulphide film possibly hinders the leaching of gold.

   Peroxides: Although a calcium ion has no effect on gold dissolution, at pH > 11.5 it can retard the gold cyanidation. Solutions kept alkaline by $Ca(OH)_2$, when compared with others at the same pH kept alkaline with KOH, hinder the leaching in the case of lime, as shown in Figure 3.9b. The decrease is presumably due to the formation of calcium peroxide (as shown in Equation 3.30) onto the gold surface, which prevents the reaction with cyanide.

$$Ca(OH)_2 + H_2O_2 = CaO_2 + 2H_2O \tag{3.30}$$

Oxides Ozone: The addition of ozone in cyanide solution decreases the leaching rate. Apparently, a layer of gold oxide that appears red brick in colour is responsible for the retarding effect. The oxidation of potassium cyanide to cyanate is also possible with ozone.

3.2.2.10 Effect of Flotation Reagents on Cyanidation

Froth-flotation is a well-established technique for concentration of gold-bearing ores prior to cyanidation. Thiol-type collectors (like potassium ethyl xanthates, and dithiophosphates) are usually employed for flotation. If the flotation products are not washed well, the presence of small amounts of thiol reagents has a negative effect on cyanidation. The poisonous effect of thiols increases with their concentration and carbon chain, which can

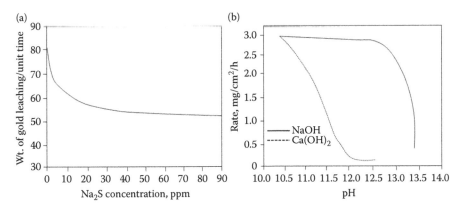

FIGURE 3.9
Leaching rate of gold in cyanide solution as a function of (a) Na_2S concentration, and (b) increasing pH by addition of different alkalis, $Ca(OH)_2$ and NaOH.

be compensated by an increased cyanide concentration (Ashurst and Finkelstein, 1970). Two different mechanisms have been proposed for this:

i. The adsorption of thiols onto mineral surfaces makes them hydrophobic and thus resistant to the diffusion of aqueous solvent. Hydrophobic particles flocculate and accumulate at the liquid-air surface.

ii. Alternatively, the adsorbed collector passivates gold surfaces. Cationic reagents used for pyrite flotation have an adverse effect on cyanidation by forming reddish film of gold-xanthate.

The rate of gold leaching decreases with the addition of as little as 0.4 ppm of potassium ethyl xanthate; however, this retardation may be largely reduced in the presence of certain fine ground silicate minerals. Other surface-active reagents such as ketones, ethers, and alcohols present at higher concentrations retard gold leaching by competing with oxygen for sites at the gold surface.

3.2.3 Intensive Cyanidation

Intensive cyanidation is an appropriate technique for extracting gold from higher concentrates, using a higher lixiviant concentration at an elevated temperature, and sometimes pressure as well (Figure 3.10). It is potent for a gravity concentrate bearing the coarse gold therein. For the purposeful enhancement of leaching kinetics, oxygen can be provided in either the form of air, oxygen, or a combination of the two (Marsden and House, 1992).

Using intensive cyanidation reactors (ICR) for gold leaching, an intensive cyanidation process requires the most favourable volume for accurate residence time for reaction with a sufficient agitation speed able to do mass transfer of gold at a higher rate. A higher lixiviant concentration optimizes leaching of gravity concentrated ore as this factor turns into the

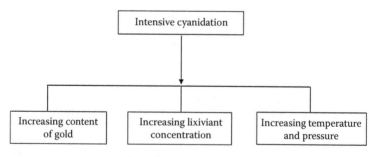

FIGURE 3.10
Application of intensive cyanidation by three ways.

rate-limiting step. The closed tank reactor of ICR ensures a great protection factor at the time of eventuality to prevent any accident, as well as environmental contamination. An ICR works in a cyclic manner using three successive tanks for filling and chemical dosing, followed by a fixed retention time and then solution discharge. Each of the tanks successively rotates to perform its function. When one of the tanks undergoes complete discharge of the leach liquor, others can continue the operation.

Intensive cyanidation has been found to be a better process than conventional techniques, as shown in Table 3.3 (Longley et al., 2003). Notably, the following techniques can be employed to enrich the gold contents of the gold-bearing ores prior to cyanide leaching of the concentrates.

3.2.3.1 Gravity Concentration Technique

The primary principle of gravity concentration is that the gold enclosed within the ore bodies has higher specific gravity than other rock minerals. Elemental gold has the specific gravity of 19.3 while the characteristic ore's gravity is ~2.6. The concentrating strategy starts movement in gold and other rock particles to separate the heavier pieces from the lighter pieces of ore. Panning is the oldest technique used in gold concentration from riverbeds, where the light weight gangue materials are separated from the top of the pan whilst leaving the heavy gold particles on the base (Eugene and Mujumdar, 2009).

TABLE 3.3

Merits and Demerits of the Intensive Cyanidation Process

Intensive cyanidation by using gold concentration techniques	Intensive cyanidation by using higher cyanide concentration	Intensive cyanidation by using higher temperature and pressure
Advantages of the Process		
Higher gold extraction	Higher gold extraction	Higher gold extraction
No extreme use of chemicals	No need of concentration unit	No need of concentration unit
Less operating cost to detoxify chemicals in waste	Not much need of optimization of leaching parameters except cyanide concentration	Less operating cost to detoxify chemicals in waste
No high need of optimization of leaching parameter		
Disadvantages of the Process		
Use of extra concentration units in plant	Extreme use of chemicals	Difficulty to attain very high temperature and pressure
High energy consumption	More arrangement for detoxification of CN waste	Dissolution of other metals due to higher cyanide level
	Specific reactor needs for high CN concentration	High energy cost
	Dissolution of other metals due to higher cyanide level	

3.2.3.2 Froth-Flotation Technique

This technique generates a concentrate by the use of chemical agents under intense stirring and air sparging, producing the mineral-rich concentrate with the froth. The addition of particular chemical agents is necessary either to float lighter minerals or to lower the flotation of heavier minerals. Flotation technique is more effective for gold bearing sulphide minerals; in contrast, the extremely oxidized ores usually do not respond well to this technique. Reagents used for flotation are commonly nontoxic, lowering the disposal cost of tailings generated during the process. Due to a higher concentration of metals, the flotation concentrates are suitable to process via smelting for recovering the precious metals. (Eugene and Mujumdar, 2009).

After concentrating the gold, the concentrate is subjected to cyanide leaching for a better extraction of gold (Eugene and Mujumdar, 2009). The intensive cyanidation has given a more prominent result than the conventional method, with several thousand-folds higher recovery (as shown in Table 3.4).

In order to enhance the leaching efficiency of gold, new reactors like the ACACIA reactor and G–ko reactor are used extensively for a higher yield of gold, >95% (Sandstrom et al., 2005), thus reducing the operation cost of cyanidation. The ACACIA reactor is also known as the ConSep ACACIA reactor. This procedure involves upflow of fluidized reactors, which produces a perfect interaction of the pulp with a maximized rate of gold leaching (Campbell and Watson, 2003). No mechanical stirring is required. The Inline Leach Reactor (ILR) or G–ko reactor can be used for either mode of operation: continuous or batch. Batch ILR is used to process high grade gold concentrates containing 1,000–20,000 g/t gold; whereas, continuous ILR is suitable for processing low grade gold concentrates (containing >50 g/t gold). Flexible application of the intensive cyanidation, such as 20 ppm dissolved oxygen, pH 13.5, 20 g/L cyanide concentration, and contact time from 6 to 24 h, can be employed (Gray et al., 2000).

TABLE 3.4

Comparison between Conditions of Intensive Gold Cyanidation and Conventional Gold Cyanidation

Sl. No.	Operating Conditions	Conventional Cyanidation Results	Intensive Cyanidation Results
1	Gold, g/t	2	20,000
2	Oxygen, ppm	8	15
3	Cyanide concentration, ppm	500	20,000
4	Leaching rate, g/h/t	0.1	1000

Source: Longley et al. (2003).

3.2.4 Gold Recovery from Cyanide Pregnant/Leach Liquor

3.2.4.1 *Zinc Cementation*

Cementation is a reductive precipitation of gold, in which the solubilized species of gold are reduced to metallic form on another metal surface lower in the galvanic series then gold. A common example of gold cementation with iron can be understood by the following equation:

$$Au^{3+} + Fe^0 = Au^0 + Fe^{3+} \qquad (3.31)$$

The electrochemical order of metals in a KCN solution indicates the sequence (from positive to negative) as: $Mg > Al > Zn > Cu > Au > Ag > Hg > Pb > Fe > Pt$. These metals follow the trend to dissolve more easily than the metals in their right form, and displace those metals by precipitation from the solution. Following the sequence, aluminium will displace gold and silver more readily than the zinc. First, MacArthur used zinc shavings in the 1890s for cementing the gold from a cyanide solution at an industrial scale. Later in 1900, C. W. Merrill introduced zinc dust, achieving more efficient recovery of gold (Habashi, 1987). The addition of zinc also evolved hydrogen gas that can contribute to gold precipitation; however, gold does not precipitate by hydrogen at atmospheric pressure. The reactions take place as shown in the following equations:

$$2Au(CN)_2^- + Zn^0 = 2Au^0 + Zn(CN)_4^{2-} \qquad (3.32)$$

$$2Au(CN)_2^- + Zn^0 + 3OH^- = 2Au^0 + HZnO_2^- + 4CN^- + H_2O \qquad (3.33)$$

$$Zn^0 + 4CN^- + 2H_2O = Zn(CN)_4^{2-} + 2OH^- + H_2 \qquad (3.34)$$

$$Au(CN)_2^- + H_2 = Au^0 + 2CN^- + 2H^+ \qquad (3.35)$$

Barin et al. (1980) have proposed an overall chemical reaction by considering the hydrogen evolution in gold cementation by zinc:

$$Au(CN)_2^- + Zn^0 + H_2O + 2CN^- = Au^0 + Zn(CN)_4^{2-} + OH^- + \frac{1}{2}H_2 \qquad (3.36)$$

Cementation is a heterogeneous redox reaction controlled by the rate by which aurocyanide and cyanide ions get transferred to the zinc surface (Finkelstein, 1972; Nicol, 1979). The reductive precipitation of gold by zinc on an industrial scale was further improved by introducing oxygen instead of air to the leach liquor.

3.2.4.1.1 *Effect of Solution Composition*

A minimum of 0.1–1.7 g/L NaCN concentration is critical for the gold cementation process (Nicol et al., 1979; Barin et al., 1980). The concentration of gold itself has a direct influence on the cementation rate that is essentially a first-order reaction controlled by the transfer rate of gold-cyanide ions. Although a change in solution pH (in range from 9 to 12) has no appreciable effect on cementation, a higher pH may lead to form intermediate hydroxides, which can retard or sometimes stop the cementation process. Finkelstein (1972) reported that the anions (like sulphate, sulphide, thiosulfate, and ferrocyanide) might reduce gold precipitation yield by 1%–2% from 10^{-3} M cyanide solutions. The free sulphates may precipitate as gypsum to reduce the reactivity in the cementation process. Nicol et al. (1979) found that sulphide ions can passivate the zinc surface even at lower concentrations of 1×10^{-4} M. For an efficient cementation recovery of gold, the leach liquor should not contain >5 ppm suspended particles and >1 ppm dissolved oxygen therein with a free cyanide concentration >0.035 M. The solution pH ideally should be in the range of 9–11 containing adequate lead nitrate (in the ratio of 1:1 with gold) and would require 5–12 parts of zinc for each part of gold to be cemented.

3.2.4.1.2 *The Merrill-Crowe Process*

Instead of using zinc shavings for gold cementation, the Merrill-Crowe process consists of four basic steps (Crowe, 1918):

   i. Clarification of the pregnant cyanide solution,
  ii. Deaeration,
 iii. Addition of zinc powder and lead salts, and
  iv. Recovery of zinc-gold precipitates.

The prime advantage of using this process over CIP arises in the presence of a high silver to gold ratio; a ratio above 4:1 commonly favours the Merrill-Crowe process.

Obtainment of a clarified pregnant solution (leach liquor) is the most vital factor; hence, the cloudy solution after a counter-current decantation is sent to a storage tank for settling of suspended particles. The entire process is presented in Figure 3.11. Notably, a complete precipitation of gold (and silver) is achieved by the removal of dissolved oxygen in the Crowe vacuum tower. Splash plates and cascade trays of the tower contribute to increase the surface area of solution; thus a complete deaeration is possible by applying the vacuum. The hydrogen evolution during cementation nullifies the effects of any traces of oxygen remaining in solution. Soon after clarification and deaeration of the pregnant solution, a continuous addition of zinc dust is performed to precipitate the gold without interruption and exposure to air. A zinc feeder introduces zinc dust as the solution flows to the precipitation filters where pressure candle filters are

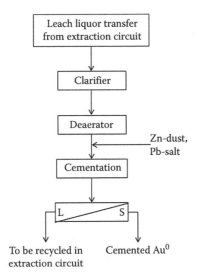

FIGURE 3.11
A process flow-sheet of gold recovery from cyanide solution via cementation using zinc-dust.

used to filter the zinc-gold slime. The pregnant leach solution has to perco-
late through the fine zinc layers, creating an extensive surface for a solid-
liquid precipitation. The resulting precipitates of metals are then mixed
and blended with flux to smelt, and thus collected bars are then subjected
for refining. The refining process usually depends on other coexisting
metals like Cu, Pt, Ag, etc.

Alternatively, aluminium can be used with NaOH solution, which is essential for the precipitation reaction as follows:

$$Al^0 = Al^{3+} + 3e^- \tag{3.37}$$

$$Al^{3+} + 3OH^- = Al(OH)_3 \tag{3.38}$$

$$Al(OH)_3 + Na^+ + OH^- = AlO_2^- + Na^+ + 2H_2O \tag{3.39}$$

$$3Au^+ + 3e^- = 3Au^0 \tag{3.40}$$

Or, the overall reaction can be written as:

$$Al^0 + 4OH^- + Na^+ + 3Au^+ = 3Au^0 + AlO_2^- + Na^+ + 2H_2O \tag{3.41}$$

However, the attempts to use aluminium as an alternate have been encum-
bered by difficulties in filtration (due to formation of calcium aluminates)
and smelting of the precipitate. In a sodium regime, the formation of soluble
aluminate proceeds the cementation process without hindrance created by

the surface films. Notably, the soluble reducing reagents (H_2S, SO_2, $NaSO_3$, and $FeSO_4$) used at a commercial level for cementing gold from chloride solutions cannot be used to quantitative precipitation of gold from the cyanide solutions.

3.2.4.2 Carbon Adsorption

Application of carbon in gold recovery is as important as the cyanidation for dissolving gold into solution. A typical carbon adsorption process for gold recovery is shown in Figure 3.12. During the deregulation of the gold price in the mid-1970s, the advent of carbon adsorption process highly impacted the economics of gold metallurgy. The adsorption mechanism of gold onto a carbon surface can be explained as:

- The adsorption of anionic gold-cyanide complex is held by electrostatic attraction or van der Waals forces.
- The gold-cyanide complex is altered to another form and precipitated in reduced form onto the carbon surface.

By replacing the Merrill–Crowe process, carbon-in-pulp (CIP), carbon-in-leach (CIL), and carbon-in-column (CIC) processes allowed the treatment of low-grade (high-clay) ores at lower capital and operating costs with higher yield. A comparison between the Merrill-Crowe and carbon adsorption processes is summarized in Table 3.5, and the details of the carbon adsorption process are given below.

- The CIP is a sequential leach and adsorption technique of gold recovery from the cyanide solution of liberated gold. During leaching pulp flows through agitation tanks for cyanidation, and then the leach (pregnant) liquor flows through agitation tanks for adsorption of gold onto activated carbon. Due to the difference in particle size,

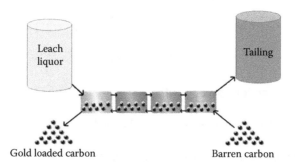

FIGURE 3.12
A typical presentation of the carbon adsorption process for gold recovery during cyanidation of gold bearing ores.

TABLE 3.5

Comparison of Carbon Adsorption Process with the Merrill-Crowe Cementation Process

Carbon Adsorption Process	
Advantages	Pretreatment of leach liquor is not needed
	Efficient recovery handling the carbonaceous ores
	Up to 99.9% recovery of all the soluble gold in cyanide solution
Disadvantages	Loss of gold with fine carbon, CIP needs more cyanide in system
	Labor intensive regeneration and elution are more expansive in practice than cementation
Merrill-Crowe Process	
Advantages	Capex, labor and maintenance cost are lower
	Can handle large Ag-to-Au ratio in leach liquor
Disadvantages	Pretreatment of leach liquor is needed prior to cementation
	Sensitive to interfering ions in the leach liquor

a screening is carried out to separate the barren pulp from gold-loaded carbon. In practice, a series of tanks (usually 5–6 tanks) are used to contact the carbon and leach slurry.

- The CIL is a simultaneous process of gold leaching (by cyanidation) and adsorption onto activated carbon, mainly applicable for processing carbonaceous gold ores. A simultaneous leaching-adsorption process helps to minimize the problem of preg-robbing.

- The CIC consists of a series of fluidized bed carbon columns in which solution flows in an up-flow direction, mostly used for gold recovery from heap leach solution. The ability to process the solution containing 2–3% wt. solids is advantageous for CIC.

Despite CIP becoming popular in the late 1970s, Davis (1880) and Johnson (1894) received patents a century ago for using the wood charcoal from chlorine and cyanide leach solutions, respectively. In 1916, Yuanmi Gold Mine in Western Australia replaced zinc precipitation by charcoal from which the adsorbed gold was typically recovered by a simple process of burning charcoal. A decade later, Gross and Scott of the U.S. Bureau of Mines published an extensive report in 1927 on recovering gold and silver from cyanide solution including parametric investigation on elution (Gross and Scott, 1927). Sodium sulphide, sodium cyanide, and sodium hydroxide were identified as the aided elution, but an efficient elution remained a serious impediment to wider acceptance of this process. In 1952, the Carlton Mill in Colorado introduced the first CIP flowsheet (Fast, 1988), employing ammonia as the complexation agent for gold elution. The use of sodium sulphide and caustic cyanide in elution became the first successful commercial elution practice and was known as the Zadra process (Zadra, 1950; Zadra et al., 1952). The installation of the Zadra elution process at CIP-using the Homestake Mine in South Dakota (Fast, 1988) is the modern CIP process now in place. Since the year 2000, several hundred plants are operating the modern CIP.

In modern CIP, the majority of activated carbon used to recover the gold from cyanide solution is either granular coconut-shell carbon or peat-based extruded carbon. Gold-loading kinetics (activity) and loading capacity, elution kinetics, level of gold elution, strength and abrasion resistance, particle-size distribution, and wet density are the criteria on which the CIP operation depends. Due consideration is therefore required on the physico-chemical properties of virgin carbon for their selection to be employed in CIP. Notably, with the standard of American Society for Testing and Materials (ASTM International) most of the activated-carbon manufacturers have their own in-house testing procedures.

Activated carbon is a heterogeneous material. The softer carbon has higher activity; hence, the loss of softer carbon due to attrition is the most active portion. Numerous methodologies to monitor the adsorption rate of gold onto carbon from the alkaline cyanide solution have been proposed (Avraamides, 1989). There is a variance in equipment used, solution composition, carbon dosage in pulp, and the particle size of carbon; but all of them involve contacting a known mass of carbon to the solution of a known gold concentration, solution analysis on a regular time-interval, and thus determining the activity of carbon with gold loading/adsorption data. One of the common phenomena observed for sorption kinetics of data often fits the rate equation well for the initial stage of adsorption. Later, when carbon loading becomes excessive the rate constant declines to fit the rate equation (La Brooy et al., 1986). A sample of virgin carbon is often used as a control with a relative activity monitor of industrial carbon in use. McArthur et al. (1887) suggested the activity of carbon should be calculated after a period of attrition. Since the loss of more active component occurs with attrition, dropping the activity of carbon, and thus collected data, would be more representative of an industrial operation. Follis (1992) advocated to determine the carbon activity by volume instead of measuring by mass. To obtain the reliable results, activity tests should be carried out in industrial process water instead of using distilled water. Approximately 30%–50% slower kinetics has been found in hypersaline solutions in comparison to distilled water (La Brooy et al., 1991).

3.2.4.2.1 Gold-Loading Capacity

The loading capacity of an activated carbon is determined from an adsorption isotherm that can be defined as the equilibrium loading onto carbon in contact with 1 mg/L of gold solution. In most cases (the Parker Centre method, Mint– method, Norit method, Anglo American Research Laboratories method), the varying masses of pulverized carbon are contacted with a gold cyanide solution for over 20 h (Davidson et al., 1982; AMIRA Project 83/P173A, 1987; Shipman, 1994; Osei-Agyemang et al., 2015).

A different approach has been used in which carbon is contacted for 1 h with 1 L of 10 mg/L gold solution. After analysing the residual gold in solution, gold adsorbed onto carbon is calculated as G1. Then the recovered carbon is put into a new batch of 10 mg/L gold solution for an additional 1 h to calculate G2. With at least 11 repetitions, the cumulative gold loading is plotted to show the increase in the gold loading on the carbon with increasing numbers of solution contacts. A higher loading on carbon is advantageous to decreasing the plant size required for elution-regeneration operation (Bailey, 1987).

3.2.4.2.2 Influential Factors Affecting the Gold Adsorption

Effect of Cyanide Concentration: A higher concentration of free cyanide has a detrimental effect on gold adsorption with carbon. However, the free cyanide in solution prevents the co-adsorption of copper; hence, a solution of free cyanide is recommended in carbon adsorption of gold from a leach liquor of copper-bearing gold ores.

Effect of Coexisting Ions: The effect of coexisting metals and the ionic strength of the solution on carbon loading is significant. The degree of gold adsorption from cyanide solution depends on the presence of cations in the solution (Davidson, 1974). Calcium aurocyanide is strongly adsorbed with carbon following the series: $Ca^{2+} > Mg^{2+} > H^+ > Li^+ > Na^+ > K^+$. Loading of gold in carbon from a deionized water is low.

Effect of pH: Depending on the solution pH, the zeta potential values of carbon change from positive to negative, and hence the adsorption b›aviour of carbon for adsorbing the protons and hydroxyl ions changes as well with respect to the pH. It has been found that the equilibrium loading of gold onto carbon increases with lower pH. The influences of Ca^{2+} and OH^- ions are antithetic in practice. A high concentration of Ca^{2+} enhances gold adsorption, while a high pH decreases the adsorption loading of gold.

Effect of Particle Size: The adsorption rate is significantly affected by the particle size of carbon. Woollacott and Erasmus (1992) suggested that the gold loading is distributed among the carbon particles. With smaller particle size, there is a higher adsorption rate due to a larger surface area for adsorption. Although the size range of particles is relatively small in a circuit, it is nevertheless significant. For the same contact time, smaller particles load higher gold than that of the larger particles, and hence, the loading of gold on an individual particle will be distributed.

3.2.4.3 Resin Ion-Exchange

The use of resin ion-exchange technique as an alteration of carbon adsorption was conceived in the late 1940s then after attempts have been made to apply on commercial scale. However, only Russia has extensively used resins in gold recovery from the cyanide solutions; first at Muruntau gold deposit in western Uzb–istan. Like carbon adsorption, the resin-in-pulp

(RIP) and resin-in-leach (RIL) are principally similar, but they are not exactly the same. In carbon adsorption, the aurocyanide complex follows adsorption of reduced gold onto stacked layers of a carbon matrix. However, the structure of ion exchange resins contains the desirable functional group with a similar charge, which can electrostatically attract the gold anions and exchange with them (Palmer, 1986). After eluting gold from activated carbon, a thermal regeneration is required before reintroduction of carbon to the adsorption circuit. In contrast, regeneration is usually performed simultaneously while eluting the gold from loaded resin. Nevertheless, the competitive sorption of unwanted anions and subsequent elution with gold has made it less attractive to the carbon adsorption. The difficulties in separating resins from pulps and their poor resistance to abrasion are also disadvantageous. On the other hand, resins can have several advantages over using carbon, such as:

- Faster kinetics and higher equilibrium loadings
- Lower temperature and pressure of elution
- No requirement for thermal reactivation
- Less sensitive to fouling and poisoning

In application, the weak-base resins with lower pK_a value need pH adjustment for adsorption of aurocyanide complex (Riveros and Cooper, 1987); however, strong-base resins can operate over a broad pH range, including the optimal pH range of the cyanidation process (in between 10 and 11). A higher loading capacity therefore requires a smaller inventory of resins, but it faces more major difficulties in elution than the weak-base resin. Elution of gold from loaded resins is being carried out by various reagents (like sodium hydroxide, sodium perchlorate, ammonium thiocyanate, zinc cyanide, acidic thiourea, and dimethylformamide). Notably, regeneration is not required while doing elution with NaOH solution; however, ammonium thiocyanate is reported to elute strong-base resins effectively, and it allows for relatively easy regeneration of the resin before recycling. The addition of sodium hydroxide to thiocyanate allows zinc, iron, and copper to be selectively eluted from the resins. A flowsheet of gold recovery by strong-base resin with regeneration is shown in Figure 3.13.

3.2.4.4 Electrowinning of Gold from Cyanide Solutions

The enriched solution of gold after elution can be transferred to electrowin the metals (including Cu, Pt, and Ag) deposition on cathode. The cathode is subsequently treated with H_2SO_4 for removal of metal impurities prior to melt and collection of the Dore. The casted Dore then undergo electro-refining to obtain pure gold, separating from the Ag and Pt. The electro-chemical reaction for electrowinnig of gold can be understood as below:

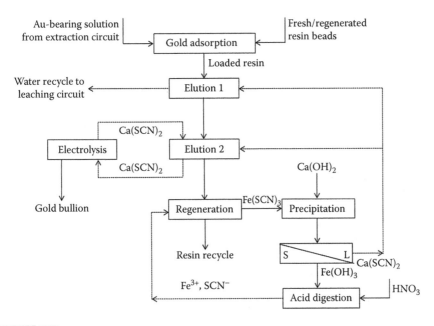

FIGURE 3.13
An adsorption-elution-resin regeneration scheme for recovering the gold from cyanide leach liquor.

$$4OH^- = O_2 + 2H_2O + 4e^- \tag{3.42}$$

$$2e^- + 2H_2O = H_2 + 2OH^- \tag{3.43}$$

$$KAu(CN)_2 = K^+ + Au(CN)_2^- \tag{3.44}$$

$$Au(CN)_2^- = Au^+ + CN^- \tag{3.45}$$

$$e^- + Au^+ = Au^0 \tag{3.46}$$

The cathode predominantly attracts positive ions near its surface that can be referred to as the Helmholtz double layer (Wilkinson, 1986). Whereas, the negatively charged $Au(CN)_2^-$ approaches this layer and becomes polarized in the electric field of the cathode. Thus, ligand's distribution around the metal is distorted and diffusion of metal complex into the Helmholtz layer is assisted to break the complex, releasing the positively charged metal cation to be deposited onto the cathode. The stability constant of aurocyanide complex can be given as (Wilkinson, 1986):

$$\frac{\left[Au(CN)_2^-\right]}{[Au^+][CN^-]^2} = 10^{38.3} \tag{3.47}$$

Hence, the extremely low concentration of [Au⁺] can be considered as:

$$[Au^+] = \frac{\left[Au(CN)_2^-\right]}{[CN^-]^2}.10^{-38.3} \tag{3.48}$$

Therefore, the rate of gold deposition from aurocyanide solution is driven by the polarization of $Au(CN)_2^-$ ions, which approach cathode surfaces and are distorted as previously described. Although the system can be operated under alkaline or neutral conditions, the concentration of [Au⁺] increases in acidic pH (3.1–7.0) due to the equilibrium of reaction and is affected by the formation of HCN (Wilkinson, 1986). It is known that the aurocyanide solution is stable without evolution of HCN, down to a pH of 3.1 (Fisher and Weimer, 1964). As the current density and cathode current efficiency increase with gold concentration, high gold containing solutions are recommended for elecrowinning. Some typical bath compositions disclosed by Fisher and Weimer (1964) are summarized in Table 3.6.

The electrolytic recovery of gold from impregnated cyanide solutions was first applied with the Siemens-Halske electrolytic method (Adamson, 1972). In this process, gold was electrolytically deposited onto lead foil cathodes, which were removed periodically, melted into ingots of lead, and processed for gold recovery. Gold electrowinning from pregnant cyanide solutions has two major inherent advantages over the chemical reduction (precipitation) of gold: (i) no addition of chemical reagents leads to recycling of the cyanide solution in the extraction process and (ii) yield of high pure gold. But a low gold in pregnant solution is a poor electrolyte, causing low current density with a slow deposition rate of gold. An ideal current density and current efficiency for gold electrowinning from cyanide solutions is proportional to the amount of gold in the electrolyte. The electrowinning of gold from pregnant mill solutions therefore copes with a low current density. In some cases, a direct electrowinning of gold from heap leach (pregnant solution) liquor has also been proposed (Eisele et al., 1986). However, the low gold contents (0.75–2.0 ppm Au) are not attractive to achieve the ideal current density and efficiency of gold deposition.

TABLE 3.6

Typical Electrolytic Bath Composition for Achieving an Efficient Electrowinning Yield of Gold from the Cyanide Solution

Metal/Ligands	Concentrations in Electrolytic Bath, g/L		
Gold, as cyanide complex	2.1	8.4	10.0
Potassium cyanide	15.0	11.0	12.0
Disodium phosphate	4.0	–	–
Alkali carbonate	Can Vary	Can Vary	Can Vary

3.2.5 Toxicity of Cyanide

For cyanidation, three types of cyanides mainly are considered; those are: (i) free cyanides, (ii) weak acid dissociable cyanides (WADs), and (iii) strong acid dissociable cyanides (SADs). However, the German Standard Methods only differentiates between the releasable cyanides (hydrocyanic acid, their alkali and alkaline earth salts, including cyanide complexes of Cd, Cu, Ag and Zn) and the strong cyanide complexes (of iron, cobalt, nickel and gold). Hence, the term total cyanides includes free cyanides, WADs, SADs, cyanate ion, OCN^-, thiocyanate ions, $SCNO^-$ and cyanogen chloride, and ClCN (Oelsner et al., 2001). The chemical name HCN stands for hydrogen cyanide and includes hydrogen cyanide, hydrocyanic acid, and prussic acid that can exist depending on the pH and redox potential either in the free or complex form. As the stability of cyanide compounds increases, the ability to release or form free cyanide in solution decreases. This is advantageous to halt the cyanide mobility but disadvantageous in the treatment/remediation of effluent as the cyanide ions form stable ferrous $[Fe(CN)_6^{4-}]$ and ferric $[Fe(CN)_6^{3-}]$ cyano-complexes.

Approximately 13%–20% of globally produced hydrogen cyanide is used in gold mining for heap leaching (Logsdon et al., 1999). The highest exposure of cyanide to an external environment is in mining, especially following spills from mines (Ok Tedt mine Papua New Guinea, Eldorado gold mine in Esme Kisladag, 1999 Turkey (Eisler and Wiemeyer, 2004) Baia Mare, Romania (Dzombac and Ghosh, 2006). Although the stoichiometry ratio of cyanide to gold is approximately 3–4 g cyanide to one-ton ore (Carrillo-Pedroza and Soria-Aguilar, 2001), in practice 200–300 g cyanide is needed for each tonne of ore. The major loss of cyanide is via the side reactions, which result in formation of various cyanide complexes. To control such losses, the International Cyanide Management Code has imposed a limit of 50 ppm cyanide solution for gold mining (Riani et al., 2007).

When cyanides are discharged into the environment, their ability to undergo a number of processes viz dissolution, adsorption, precipitation, (bio)oxidation, and biodegradation, along with a variety of metal complexes, makes it difficult to trace their path in the soil, water, or air. These processes often occur simultaneously, depending on the prevailing physical or chemical conditions, and result in degradation or attenuation of cyanides. The toxicity of cyanides is a function of dissociation of free cyanides into the environment. Their dissociation reduces with increasing stability of the cyanide complex, from weak complexes to strong complexes. The stability constants of the cyanide complexes with lead, silver, copper, and nickel are lesser than ion complexes, releasing HCN more easily and therefore are more toxic (Oelsner et al., 2001). The cyano-complexes of copper and zinc in WADs are insoluble in water but soluble in ammonia solution. The production of ammonia from natural attenuation would therefore result in dissolution of copper and zinc in cyanide solution. This results in an increase in bioavailability and cyanide concentration within a specific environment. Except for volatilization of the free cyanide, the released cyanides may still be available to drive other chemical processes under suitable conditions. This holds true for photolysis, precipitation, and complexation processes. Cyanide in the open environment undergoes a number of redox reactions, forming various cyanide species of different toxicity levels, especially at higher concentrations. These compounds include:

- Cyanates: They are usually formed as intermediate products via reaction of cyanide with oxidants (ozone, hypochlorite, chlorine, hydrogen peroxide) during oxidative degradation of cyanides.
- Cyanogen Chlorides: They are toxic compounds, formed as intermediates during the chlorine oxidation of cyanides to cyanates. In the presence of ammonia, another class of toxic compounds, chloramines, are formed.
- Thiocyanates: They are formed by the action of cyanide with sulphur or sulphur containing chemical species and remain present in mineral ores. They persist in acid mine drains for decades after the closing of mines.
- Cyanogens: They are produced in acidic environments when free cyanide encounters oxidants like the oxidized copper minerals; nevertheless, their formation is not expected in alkaline conditions.
- Nitrate and Ammonia: A chemical dissociation of cyanides and previously mentioned cyanide derivatives generate a high number of nitrates and ammonia as the degradation products; these are toxic to aquatic organisms.

Cyanide is a toxic substance; it can be inhaled, ingested orally (through contaminated water or food), and diffused through the skin. Cyanide prevents the intake and subsequent transportation of oxygen to the cells (Logsdon et al., 1999). Iron, which coordinates the intake and transport of oxygen to the cells, is consumed by cyanide, causing failure of the respiratory system, rapid breathing convulsions, loss of consciousness, and suffocation if there is no medical intervention. However, the body can detoxify small concentrations of cyanide to less toxic cyanate, preventing the accumulation of cyanide in the human body. Cyanide diffusion through the skin is supported by the small molecular size of HCN and the fact that cyanide dissolves readily in lipids of human bodies (Simeonova and Fishbein, 2004). Concentrations of 20–40 ppm HCN in the air are toxic; increasing the concentration up to 250 ppm (1–3 mg CN per kg body weight) causes death within minutes. The LD_{50} values representing the toxicity of cyanide and its derivatives are given in Table 3.7.

3.2.5.1 Toxicity Standards for Cyanide Disposal

To protect the water reservoirs from the toxicity of cyanide and its compounds, the contaminated water must be detoxified before its discharge. The U.S. Environmental Protection Agency (USEPA) has limited 200 ppb

TABLE 3.7

Toxicity of Cyanide and Cyanide Derivatives

Substance	B.P. (°C)	Exposure Limit	LD_{50}	Cyanide Release
Bromobenzylcyanide	Solid	–	3.5 g/g	+
Cyanamide	Solid	2 mg/g	1.0 g/g	0
Cyanide salts	Solid	5 mg/g	2.0 mg/g	+
Cyanoacetic acid	108	–	2.0 g/g	0
Cyanogen	Gas	10 ppm	13.0 mg/g	+
Cyanogen chloride	61	0.3 ppm	13.0 mg/g	+
Ferric-cyanide	Solid	–	1.6 g/g	0
Ferroc-cyanide	Solid	–	1.6 g/g	0
Hydrogen cyanide	26.5	10 ppm	0.5 mg/g	+
Malonitrile	Solid	3 ppm	6.0 mg/g	+
Methylcyano-acrylate	Liquid	2 ppm	–	+
Methyliso-cyanate	39	0.02 ppm	2 ppm	–
Nitoprusside	Solid	–	10 mg/g	+
O-Tolunitrile	204	–	0.6 g/g	0

Source: Low›ein and Moran (1975).

and 50 ppb cyanide in drinking water and for aquatic-biota, respectively (Gurbuz et al., 2004). The National Primary Drinking Water Regulations have proposed a permissible limit for cyanide of 0.2 mg/L in effluent and 0.01 mg/L in the drinking water as guidelines (EPA, 2017). The Swiss and German standards for regulation of cyanide are 0.01 mg/L for drinking or surface water and 0.5 mg/L for effluents. In Mexico the disposal regulation for cyanide is 0.2 mg/L. The minimal national standard (MINAS) for the discharge of cyanide as stated by the India Central Pollution Control Board (CPCB) is 0.2 mg/L. In order to avoid the toxicity of cyanide, it is important to treat the industrial waste water (Dash et al., 2008).

3.2.6 Methods of Cyanide Destruction/Removal

3.2.6.1 Ion Exchange

The ion exchange is a process in which mobile ions from an external solution are exchanged for ions that are electrostatically bound to the functional groups contained within a solid matrix (IAEA, 2002). The material is constructed such that it can selectively absorb either cations or anions from a solution in contact with it, and in exchange this material releases counter ions from its ion exchange sites (IAEA, 2002). Resins are classified based on the strength (strong or weak) and charge (positive or negative) of their functional groups as these two factors determine which ions a resin can take out of a solution. The beads of strong acid cation (SAC) resins, strong base anion

(SBA) resins, and weak acid cation (WAC) resins have two components: a fixed or a mobile ion exchange site and a mobile ion.

The anions like $Cu(CN)^{2-}$, $Zn(CN)^{2-}$, $Ni(CN)_4^{2-}$, or only CN^- as free cyanide dissociate in water. A strong base resin with a functional group of quaternary ammines having a permanent positive charge can be used to extract the cyanide species from the solution. For example, the ion exchange process responsible for the adsorption of the zinc cyanide complex can be written as follows:

$$\int 2\left(N^+R_3X^-\right) + Zn(CN)_4^{2-} = \int (N^+R_3)_2 Zn(CN)_4^{2-} + 2X^- \qquad (3.49)$$

Advantages of ion exchange methods included no formation of new products, easy recovery of metals, and more versatility than other material; however, a major problem with this method has been disposal of the spent solution from the process. Currently there are many methods for treating these spent solutions. Membrane filtration and electro-dialysis are effectively used for cyanide removal from effluents. Nevertheless, short lifetime and continual replacement of membrane are major drawbacks associated with these methods (Ahmaruzzaman, 2011). Adsorption is preferred over membrane filtration due to design simplicity, insensitivity, flexibility, and low-cost adsorbents such as activated carbon, biomaterial, and synthetic material (Sorokin et al., 2001; Kim et al., 2003). In the adsorption mechanism, molecules of adsorbate are attached on the surface of the adsorbent by strong binding forces and separate from the effluent. The dilution of cyanide containing effluents neither separates nor destroys the free cyanides. It simply produces an effluent that has a minute concentration of cyanides below the discharge limits. Although it is cheap and easy to use, it does not reduce the absolute amount of cyanide in the discharge stream. Once released to the environment, cyanide can be concentrated by adsorption and/or precipitation in surface and ground waters. Hence, an electrolytic process was developed as an alternative. In this process, the electrolytic cell is placed with an alternative series of cathodes and anodes in an arrangement fitted with air spargers and flow dispersers.

3.2.6.2 Electrowinning

In electrowinning, an electric potential is applied across two electrodes immersed in an electrolyte (cyanide-metal-complex bearing solution). As the electric potential is applied, the metals are stripped-off from their corresponding complexes with cyanides, leaving free cyanide ions which are attracted (due to their negative charge) to the positively charged electrode, the anode. Under appropriate electric potential conditions, the dissolved metal cations migrate to the negative electrode, the cathode, where they are reduced and deposited/plated in metallic form onto the cathode, and the negative ions (free cyanide) are oxidized at the positively charged anode (Wang et al., 2006). When an electric potential is applied to a solution containing WADs and SADs, the metals (M) dissociate from these complexes as:

$$M(CN)_x^{y-x} + ye^- = M^0 + xCN^-$$

(3.50)

Considering an electrolyte containing silver cyanide, the following equations occur:

$$CN^- + 2OH^- = CNO^- + H_2O + 2e^-$$

(3.51)

$$[Ag(CN)_2]^{2-} + 6OH^- = Ag^{2+} + 3CNO^- + 3H_2O + 5e^-$$

(3.52)

$$Ag^{2+} + 2e^- = Ag^0$$

(3.53)

$$2CNO^- + 4OH^- = 2CO_2 + N_2 + 2H_2O + 6e^-$$

(3.54)

The electrowinning reaction for sodium copper cyanide complex can be written as:

$$4\,Na_2Cu(CN)_3 + 4\,NaOH = O_2 + 2H_2O + 4\,Cu^0 + 12\,NaCN$$

(3.55)

The electrolytic process does not produce any toxic sludge that requires further treatment. It favours the metal's recovery which is unaffected by the presence of organic or metallic impurities (Sobral et al., 2002). But the mass transfer rate reduces with the reduction of the concentration gradient; therefore the rate of metal electroplating on the cathode also reduces (Sobral et al., 2002). A reduction in the plating-out of the metals from the solution implies a lower efficiency of the process. Another problem is that the CN⁻ concentration near the cathode increases significantly, retarding the reduction process and even possibly changing the order of the metal complex to a higher value that hinders the plating of metal ions on to the cathode (Volesky and Naja, 2005; Sobral et al., 2002). These problems are resolved through the application of cathodes with very large surfaces (flow-through processes) or by increasing turbulence (flow-by processes) in the solution close to the cathode (Sobral et al., 2002; Volesky and Naja, 2005). But a major problem with these cell types is a high probability of breakdown, since many small parts are involved and there is a high cost of process. While flow-through processes have advantages of continuity by bleeding the particles into a recycle stream, the complexity of the process and the cost involved have not made it attractive as a metal recovery option and consequently not for cyanide treatment.

3.2.6.3 Hydrolysis Distillation

The cyanide ion readily hydrolyses in water producing gaseous hydrogen cyanide as shown in the following equations.

$$H_2O = OH^- + H^+ \tag{3.56}$$

$$H^+ + CN^- = HCN \tag{3.57}$$

$$CN^- + H_2O = HCN \uparrow + OH^- \tag{3.58}$$

In the case of iron-complex, the metals get hydrolysis precipitation.

$$3OH^- + Fe^{3+} = Fe(OH)_3 \downarrow \tag{3.59}$$

In alkaline environments with increased temperatures, free cyanide reacts with water to give formate and ammonia as shown in the following equation.

$$CN^- + 2H_2O = HCOO^- + NH_3 \tag{3.60}$$

Hydrogen cyanide undergoes complete hydrolysis in water, and the rate of hydrolysis is temperature dependent accompanied by an increase in the redox potential (Oelsner et al., 2001). The product of hydrolysis is pH dependent; at low pH ammonium formate (Equation 3.61) is produced while at high pH formic acid (Equation 3.62) is produced.

$$HCN + 2H_2O = HCOONH_4 \tag{3.61}$$

$$HCN + 2H_2O = NH_3 + HCOOH \tag{3.62}$$

The aqueous HCN can volatize in gaseous form, which has a lower vapour pressure than water and therefore goes into the gaseous phase much faster than water as:

$$HCN_{(aq)} = HCN_{(g)} \tag{3.63}$$

At a normal room temperature (~298 K), the HCN gas has much higher vapor pressure (100 KPa) than that of water (34 KPa). The ability of HCN to get into the gaseous phase more readily than water can be enhanced to separate it from water by elevated temperature of the solution, reducing the external pressure and by increasing the surface area of the solution agitation. Increasing the surface of the solution is important since only the hydrocyanic liquid on the surface is able to get to the gaseous phase. The treatment of cyanide by hydrolysis and evaporation in tailings ponds must be strictly regulated to avoid poisoning.

3.2.6.4 Membrane Treatment

Membrane treatment is the separation of suspended and/or dissolved solids from a solution with the use of a semi-permeable membrane. There are basically two types of membrane processes: (i) the pressure-driven and (ii) the electrically driven processes (EPRI, 1997). Pressure-driven processes utilize hydraulic pressure to force gold cyanidation discharges to pass through a membrane. As the discharge is forced through the membrane, a concentrate of impurities (ions, organic matter, or bacteria, depending on the permeability of the membrane) is retained in the feed solution. The pressure-driven processes can be further categorized based on membrane permeability. In order of decreasing permeability these processes can be classified as microfiltration (MF), ultrafiltration (UF), nanofiltration (NF), and reverse osmosis (RO). In RO, hydraulic pressure is applied to a solution containing cyanide, and water is forced out of the solution through a membrane impermeable to cyanide. This process is currently of greater interest in the field of cyanide separation/treatment than the others.

3.2.6.5 Electro Dialysis

The electro dialysis processes are driven by an electrical current or by an electric potential. The filters/membranes employed in electro dialysis are semipermeable to ions based on the charge of the ions, the electric potential applied, and on their ability to reduce the ionic content of water (Osmonics, 1997). An improved version of this process is the electro dialysis reversal (EDR) process. Periodic reversal of the direct-current driving force prevents scaling and fouling of the membrane surface as in the electro dialysis processes (Osmonics, 1997), increasing the efficiency and operational life of the membranes. The cyanide solution to be treated is placed in the cathode chamber. The negatively charged cyanide ions migrate through the membrane to the anodic chamber.

3.2.6.6 Complexation Methods

3.2.6.6.1 Acidification, Volatilization, and Recovery (AVR)

This process, also known as the Mills-Crowe process, was developed around 1930 at the Flin Flon operation in Canada. It is based on the observation that when the pH of a cyanide solution drops below 9.3, there is formation of HCN (Miller, 2003). In most AVR processes, sulfuric acid is often employed. In the AVR process the escaping hydrocyanic gas is channelled through an alkaline medium where it is captured or stripped off from the flowing gas. The chambers where the mixing of the cyanide solutions and the acids occur are sealed to prevent any escape of the HCN gas. The lower the pH falls, the higher the rate of formation and volatilization of hydrocyanic gas. At pH values ≤2, HCN is evolved from the WAD cyanide complexes (Young and Jordan, 1995).

$$M(CN)_x^{x-y} + xH^+ = x(HCN)_{(g)} + M^{y+} \qquad (3.64)$$

$$Cu\,(CN)_3^{2-} + H_2SO_4 = CuCN_{(s)} + 2HCN_{(g)} + SO_4^{2-} \qquad (3.65)$$

$$2HCN_{(g)} + Ca(OH)_2 = Ca(CN)_{2(aq)} + 2H_2O \qquad (3.66)$$

$$Ca(CN)_2 + H_2SO_4 = CaSO_4 + 2HCN \qquad (3.67)$$

Approximately 67% of copper cyanide complex can be treated in this manner, while 33% remains complexed (Davies et al., 1998). SADs and thiocyanates can also be treated in a similar manner, but the pH must be below zero, consuming high acid. For this reason, AVR processes are done at pH values between 1.5 and 2.0 resulting in only a slight alteration of SADs and thiocyanates. In such cases where lime is used to neutralize the cyanide, care should be taken to avoid using H_2SO_4 for acidification to prevent gypsum precipitation.

The liquid leaving the reaction chamber is stripped using a stream of air in a packed column. The cyanide-laden air passes through a second chamber flowing counter to a stream of alkaline solution of pH ~11.0 (Barr et al., 2007). The formation of sodium cyanide takes place as shown in the following equation:

$$HCN + NaOH = NaCN + H_2O \qquad (3.68)$$

This technology is good as the recovered cyanide can be recycled in the leaching process; also, the production of poisonous products (cyanates, thiocyanates, chloramines, ammonia, etc.) does not arise as it does in oxidation. It can be particularly good for remote locations to reduce the cost of constantly bringing in new cyanide for leaching (Miller, 2003). However, the system has a series of drawbacks including a demand of high investment for plant installation and energy for aeration. It barely achieves the cyanide waste emission standards; therefore, the product requires further treatment before final disposal. Because of the high cost of acids, the AVR process is best applied for small amounts of waste with very high cyanide concentration. Therefore, it can be best applied in conjunction with a pre-concentration stage, which may be by ion exchange or solvent extraction (Lee, 2005).

3.2.6.6.2 Flotation

Bucsh et al. (1980) demonstrated that cyanide-metal complexes can be extracted from solution by ion-precipitate flotation, if the cyanides are precipitated with an appropriate metal. Latkowska and Figa (2007) also demonstrated the flotation of free cyanides after coagulation and flocculation with appropriate reagents. The ionic ends of the quaternary amine bonded the collector to cyanide complexes while the non-ionic end attached to air bubbles providing buoyancy to float these species to the surface for collection.

$$(y-x)R_4NCl + M(CN)_x^{(y-x)} = (R_4N)_{y-x}M(CN)_{6(s)} + (y-x)Cl^- \qquad (3.69)$$

A schematic of the flotation column is shown in Figure 3.14 (Kawatra, 2002). Flotation of cyanides is only possible after formation of insoluble cyanide complexes. Flotation as a treatment method would not qualify for complete treatment of cyanide wastes as it can only partially treat WADs and the remediation of thiocyanates has not been established. It can be used as a pretreatment method before the application of more aggressive cyanide treatment methods.

3.2.6.6.3 Addition of Metal Ions

In this method, an appropriate metal salt is added into a cyanide solution with the aim of converting the cyanide species to a form either more susceptible to available remediation technologies or less toxic. This technique is suitable in the gold mining industry using the Merrill-Crowe process. With the addition of zinc salt to leach liquor, gold ions are replaced by zinc in the aurocyanide complex, precipitating pure gold and leaving residual zinc cyanide complex in solution.

Other metals capable of precipitating gold from the cyanidation leachate include aluminium, copper, and iron (Young et al., 1984); the precipitations are pH dependent. At pH > 8.5, Fe^{2+} reacts with cyanide to form the $Fe(CN)_6^{4-}$, while Fe^{3+} forms $Fe(CN)_6^{3-}$ at a pH between 5 and 8 (Latkowska and Figa, 2007). Although these compounds are very stable, they decompose in UV light. The addition of Fe^{3+} at a pH between 3 and 4 leads the reaction as:

$$3Fe(CN)_6^{4-} + 4Fe^{3+} = Fe_4[Fe(CN)_6]_{3(s)} \qquad (3.70)$$

Ferrohexacyanide anion reacts with various metal ions at a pH of between 3 and 5, as follows:

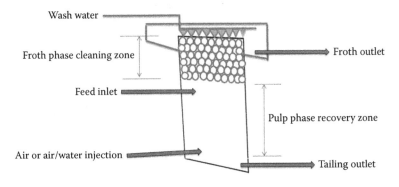

FIGURE 3.14
Schematic of a flotation column. (Modified from Kawatra, 2002).

$$Fe(CN)_6^{4-} + 2M^{2+} = M_2[Fe(CN)_6]_{(s)} \tag{3.71}$$

In a similar manner, the double salt precipitation of ferrihexacyanide occurs as:

$$2Fe(CN)_6^{3-} + 3Fe^{2+} = Fe_3[Fe(CN)_6]_{2(s)} \tag{3.72}$$

When ferrihexacyanide anions react with ferric ions, Prussian green at a pH between 3 and 4 forms as:

$$Fe(CN)_6^{3-} + Fe^{3+} = Fe_2(CN)_{6(S)} \tag{3.73}$$

3.2.6.7 Alkali Chlorination

The schematic flow of alkali chlorination process for cyanide destruction is shown in Figure 3.15. In the process, the oxidation of cyanides by chlorine occurs in two steps: (i) the conversion of cyanide to cyanogen chloride or tear gas and (ii) the hydrolysis of cyanogen chloride to cyanate as:

$$Cl_{2(g)} + CN^- = CNCl + Cl^- \tag{3.74}$$

$$CNCl + H_2O = CNO^- + Cl^- + 2H^+ \tag{3.75}$$

The proton ions produced as in Equation 3.75 neutralize lime in the solution, pushing pH downwards and away from optimum condition. Hence, an alkali must be added to compensate the pH drop. The chlorine introduced into the system does not only react with cyanides, but also with WADS

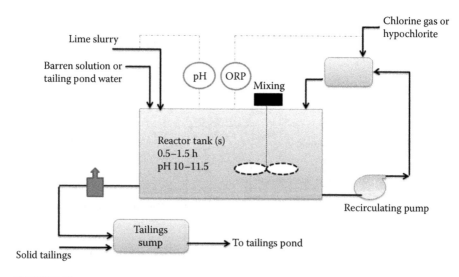

FIGURE 3.15
Alkaline chlorination of cyanide. (Modified from Dzombac and Ghosh, 2006).

and thiocyanides (Young et al., 1984). The high pH in the reaction chamber favours the formation of sulphates and the metal hydroxides formed precipitate (Young et al., 1984). In both situations, there is high consumption of lime accounting for its demand in the process by the following reactions:

$$SCN^- + 4Cl_2 + 10(OH)^- = OCN^- + SO_4^{2-} + 8Cl^- + 5H_2O \qquad (3.76)$$

$$M[(CN)_x]^{y-x} + xCl_2 + (2x+y)(OH)^- = xOCN^- + 2xCl^- + M(OH)_y + xH_2O$$
$$(3.77)$$

The chemical composition of the effluent produced from the alkaline chlorination of cyanides depends on the quantity of excess chlorine present in the reaction chamber (Botz, 1999). In the presence of a small excess of chlorine, the cyanates hydrolyzed to ammonia as follows:

$$OCN^- + 3H_2O = NH_4^+ + CHO_3^- + OH^- \qquad (3.78)$$

In a situation of a large excess of chlorine, a further oxidation of ammonia completely to nitrogen gas occurs as:

$$3Cl_2 + 2NH_4^+ = N_2 + 6Cl^- + 8H^+ \qquad (3.79)$$

Theoretically, 2.73 g chlorine is required to oxidize 1.0 g cyanide to cyanate, but in actuality it ranges from 3.0–8.0 g Cl_2. The chlorine can be provided as compressed liquid Cl_2, as a 12.5% NaOCl solution. The reaction is carried out at a pH > 10.5 to ensure cyanogen chloride is rapidly hydrolysed to cyanate. At a pH lower than 10.5, cyanogen chloride may evolve as toxic HCN gas. In this process, copper is not required as a catalyst, as it is in other processes. Upon completion of the reaction, metals previously complexed with cyanide are precipitated as hydroxides. The presence of chloramines inhibits the completion of the reaction, resulting in the possible appearance of insoluble metallic cyanides in the effluent (ASTI, 2007). This situation can be avoided by avoiding the breakpoint phenomenon.

3.2.6.8 Hypochlorite Oxidation of Cyanides

As a result of the observations of chlorine oxidation where hypochlorite produced by chlorine hydrolysis reaction (Equation 3.80) oxidizes cyanides to cyanogenchloride (Equation 3.81) that further hydrolyses to cyanates (Equation 3.82), a process for cyanide destruction using hypochlorite was also conceived (Young and Jordan, 1995).

$$Cl_{2(g)} + 2OH^- = Cl^- + OCl^- + H_2O \qquad (3.80)$$

$$CN^- + OCl^- + H_2O = CNCl_{(aq)} + 2OH^- \qquad (3.81)$$

$$CNCl_{(aq)} + 2OH^- = OCN^- + Cl^- + H_2O \qquad (3.82)$$

Sodium hypochlorite is commonly used for hypochlorite oxidation of cyanides (Durney, 1984); when used, its alkalinity reduces the demand for a base. Hypochlorites also react with thiocyanates and metal-cyanide complexes as shown in Equations 3.83 and 3.84, respectively.

$$SCN^- + 4OCl^- + 2OH^- = OCN^- + SO_4^{2-} + 4Cl^- + H_2 \qquad (3.83)$$

$$M(CN)_x^{y-x} + xOCl^- + yOH^- = xOCN^- + xCl^- + M(OH)_y \qquad (3.84)$$

The excess hypochlorite in solution reacts with cyanates to give nitrogen and carbon dioxide gages:

$$2OCN^- + 3OCl^- + H_2O = N_{2(g)} + 2CO_{2(g)} + 3Cl^- + 2OH^- \qquad (3.85)$$

The oxidation of sodium cyanide with hypochlorite can be written as:

$$5NaOCl + H_2O + 2NaCN = 2NaHCO_3 + N_2 + 5NaCl \qquad (3.86)$$

Hypochlorite has 50% the neutralizing capacity of lime. This makes the use of hypochlorite more economically attractive than chlorine application, especially when dealing with small waste amounts.

3.2.6.9 Hydrogen Peroxide Process

Hydrogen peroxide, a stronger oxidant than oxygen, is preferred as an oxidant for treating cyanide due to its being less expensive, water soluble, and easy to handle and store (Young and Jordan, 1995). Therefore, this method has long been applied, mainly for treating cyanide waste batch solutions where the treatment of slurries has been found to be difficult (Logsdon et al., 1999). Initial attempts to use H_2O_2 showed a consumption far above the reaction stoichiometry.

The advantages of this process include the non-production of environmentally harmful by-products due to the reaction products being soluble salts (Iordache et al., 2003). This process oxidizes free cyanides and WAD cyanides to cyanates. The metals released from the WAD cyanides are precipitated as metal hydroxides while stable iron-cyanide complexes are removed as insoluble copper iron-cyanide complexes. The peroxide oxidation of WAD cyanide complexes occurs as follows:

$$Me(CN)_4^{-2} + 4H_2O_2 + 2HO^- = Me(OH)_2 + 4OCN^- + 4H_2O \qquad (3.87)$$

Because this method does not treat SADs it is also referred to as a selective detoxification process (Griffiths et al., 1987). Depending on the pH in the reaction chamber, the cyanates slowly hydrolyse further, giving off either carbon dioxide and ammonium or a carbonate and ammonia, as shown in the following reactions:

$$CNO^- + 2H^+ + H_2O = CO_2 + NH_4^+ \tag{3.88}$$

$$CNO^- + OH^- + H_2O = CO_3^{2-} + NH_3 \tag{3.89}$$

3.2.6.10 Ozone Treatment of Cyanides

The oxidizing power of ozone is five-folds higher than oxygen and about twice more than chlorine, making it more reactive (Oxomax, 2005). Cyanide oxidation to cyanate by ozone occurs at pH 9–10. It is a spontaneous reaction with the rate of reaction only limited by gas transferred to the liquid phase (Carrillo-Pedroza and Soria-Aguilar, 2001), either following the simple pathway (Equation 3.90) or, the catalytic pathway (Equation 3.91).

$$CN^- + O_{3(aq)} = OCN_{2(aq)}^- \tag{3.90}$$

$$3CN^- + O_{3(aq)} = 3OCN_{2(aq)}^- \tag{3.91}$$

Ozonation is effective for treating the free cyanides, WADs, and thiocyanates but not for SADs (Huiatt, 1984). Hydroxide readily decomposes ozone resulting in a slowdown of ozone oxidation of cyanide species at pH > 11.0. The oxidation of cyanides is done in the pH ranges 10–12 where the reaction rate is good and relatively constant (National Risk Management Research Laboratory, 2000).

3.2.6.11 INCO Process

In the INCO process, oxidation of cyanides (EPA, 1994) to cyanates using a mixture of sulphur dioxide and air in the presence of a catalyst at controlled pH is performed (Scott, 1984). The process treats all forms of cyanides; free cyanides and WADs are oxidized to cyanates and the metals involved in the WADs complexes are precipitated as hydroxides. The ferric cyanide complexes are reduced to the ferrous state, and precipitated as insoluble double salts of the respective metals (Terry et al., 2001). The process uses sulphur dioxide as a gas and sulphite as a salt or as a solution in combination with air as an oxidant. To obtain optimum pH for the reaction, lime is needed for pH regulation and soluble copper issued as a catalyst (Scott, 1984; Lemos et al., 2006). In the sulphur dioxide process designed by Noranda Incorporated, only pure sulphur dioxide is used. The main difference is that the Noranda process is best suited for sites with high concentration of antimony and arsenic. This process can be divided into oxidation (free cyanide and WADs), neutralization, precipitation, and copper catalysis (Equations 3.92–3.97).

$$CN^- + SO_2 + O_2 + H_2O = OCN + H_2SO_4 \tag{3.92}$$

$$Me(CN)_4^{2-} + 4SO_2 + 4O_2 + 4H_2O = 4OCN + H_2SO_4 + Me^+ \tag{3.93}$$

$$H_2SO_4 + Ca(OH)_2 = CaSO_4 \cdot 2H_2O \tag{3.94}$$

$$Me^+ + Ca(OH)_2 = Me(OH)_{2 \text{ (precipitated)}} \tag{3.95}$$

$$2Me^+ + Fe(CN)_6^2 = MeFe(CN)_{6 \text{ (precipitated)}} \tag{3.96}$$

$$SO_2 / CN_{WAD} = 46.2 \text{ g} / \text{g} \tag{3.97}$$

Copper as $CuSO_4$ plays important roles, as it catalyses the oxidation reaction and assists the precipitation of ferrocyanide as well. The free cyanide in the solution quickly complexes with $CuSO_4$ giving off Cu(CN), which is believed to be the catalyst for oxidation of CN^- to CNO^- by SO_2 and O_2 (Huiatt, 1984).

3.2.6.12 Photolytic Degradation

Photolytic reactions are characterized by a free radical mechanism initiated by interactions of photons of a particular energy level with the molecules of chemical species in the presence or absence of a catalyst (Gogate and Pandit, 2004). A direct photochemical degradation or photolysis of cyanide is possible for WADs, SADs, and for free cyanides. Cyanide photolysis has been reported as being partly responsible for the reduction of cyanide amounts in tailings ponds (Simovic and Snodgrass, 1999; Botz and Mudder, 2000). This is especially possible for iron cyanide complexes in the following equations:

$$Fe(CN)_6^{3-} + H_2O = [Fe(CN)_5 H_2O]^{2-} + CN^- \tag{3.98}$$

$$[Fe(CN)_5 H_2O]^{2-} + 6H_2O = Fe(OH)_{3(S)} + 5CN^- + 3H^+ \tag{3.99}$$

$$Fe(CN)_6^{3-} + 6H_2O = [Fe(H_2O)_6]^{3+} + 6CN^- \tag{3.100}$$

A photolytic process in the presence of ozone can also be used as in the following reactions:

$$H_2O + O_3 = 2OH^- + O_2 \qquad (3.101)$$

$$CN^- + 2OH^- = OCN^- + H_2O \qquad (3.102)$$

Because the hydroxyl radicals formed do not possess any charge they have a higher affinity for electrons than ozone and so will attack any chemicals including thiocyanates, WADs, and SADs, oxidizing them in the process. The cyanates formed are then hydrolyzed as follows:

$$OCN^- + 3H_2O = NH_4^+ + HCO^{3-} + OH^- \qquad (3.103)$$

Otherwise, the cyanate is degraded by photolytic ozonation to give bicarbonate, and one of nitrogen compound as:

$$OCN^- + 3OH^+ = HCO_3^- + \frac{1}{2}N_2 + H_2O \qquad (3.104)$$

$$OCN^- + 6OH^+ = HCO_3^- + NO_2^- + H^+ + 2H_2O \qquad (3.105)$$

$$OCN^- + 8OH^+ = HCO_3^- + NO_3^- + H^+ + 3H_2O \qquad (3.106)$$

3.2.6.13 Heterogeneous Photocatalysis

Heterogeneous photocatalysis is based on using a source of UV radiation to perform the oxidation on the stimulated surface of a semiconductor material (Aguado et al., 2002). Three reaction paths have been proposed for the oxidation of cyanides in heterogeneous photocatalysis. The first reaction path is believed to occur through the production of powerful hydroxyl radicals, which subsequently oxidizes any cyanide present in the solution. Production of the hydroxyl radicals occurs through the reduction of water adsorbed on the semiconductor surfaces or the reduction of adsorbed hydroxide by the holes (h^+) in the valence band, as shown in the following reactions:

$$\text{Conductor} = (e^- h^+)(x) \qquad (3.107)$$

$$H_2O + h^+ = OH^- + H^+ \qquad (3.108)$$

In a second reaction path, the dissolved oxygen reacts with the excited electrons in conduction band through a superoxide as an intermediate. The hydroxyl radicals produced in both previous reaction stages then react with cyanides, resulting in cyanate hydrolysis as follows:

$$OH^- + h^+ = OH^- \qquad (3.109)$$

$$O_{2(aq)} + e^- = O_2^- \tag{3.110}$$

$$2O_2^- + 2H^+ = 2OH^- + O_2 \tag{3.111}$$

$$CN^- + 2OH^- = OCN^- + H_2O \tag{3.112}$$

$$OCN^- + 3H_2O = NH_4 + + HCO_3^- + OH^- \tag{3.113}$$

The cyanate involved in reactions with the hydroxyl radical results in the formation of nitrogen gas, nitrite, and/or nitrate. The first and second cyanide oxidation paths are effective in treating thiocyanates, WADs, and SADs. Any ammonia produced during cyanate hydrolysis is oxidized. In the third reaction path, cyanide is reduced by the holes (h^+) in the valence band producing cyanyl radicals as:

$$CN^- + h^+ = CN^0 \tag{3.114}$$

The cyanyl radicals are then attacked and oxidized by the previously produced hydroxyl radicals, and hydrolyse the cyanates.

References

Adamson, R.J. 1972. *Gold Metallurgy in South Africa*. Chamber of mines of South Africa, Johannesburg, pp. 203–255.

Aghamirian, M., Yen, W.T. 2005. Mechanisms of galvanic interactions between gold and sulfide minerals in cyanide solution. *Minerals Engineering.* 18(4): 393–407.

Aguado, J., Gri–e, R.V., López-Muñoz, M., Marugán, J. 2002. Removal of cyanides in wastewater by supported Tio$_2$-based photocatalysis. *Catalysis Today.* 75: 95–102.

Ahmaruzzaman, M. 2011. Industrial wastes as low-cost potential adsorbents for the treatment of wastewater laden with heavy metals. *Advances in Colloid and Interface Science.* 166(1): 36–59.

AMIRA, P.P.A. 1987. Carbon-in-pulp gold technology. Progress report no. 4. Anglo American Research Laboratories Ltd. Determination of the platelet content of activated carbon. AARL/AAC specification test procedure.

Ashurst, K.G., Finkelstein, N.P. 1970. The influence of sulphydryl and cationic flotation reagents on the cyanidation of native gold. *Journal of Southern African Institute of Mining and Metallurgy.* 70: 243–258.

ASTI, A.S. 2007. *Cyanide Wastes*. Advanced Sensor Technologies, Inc., Orange, CA.

Avraamides, J. 1989. CIP carbons—Selection, testing and plant operations. In: Bhappu, B., Harden, R.J. (Eds), *Gold Forum on Technology and Practices-World Gold*, SME, Littleton, CO, pp. 288–292.

Bailey, P. 1987. Application of activated carbon to gold recovery (retroactive coverage). *South African Institute of Mining and Metallurgy, The Extractive Metallurgy of Gold in South Africa.* 1: 379–614.

Barin, I., Barth, H., Yaman, A. 1980. Electrochemical investigations of the kinetics of gold cementation by zinc from cyanide solutions. *Erzmetall.* 33: 399–403.

Barr, G., Willy, G., David, J., Keith, M. 2007. *The New Cesl Gold Process.* Cominco Engineering Services Ltd, Perth, WA.

Barsky, G., Swainson, S., Hedley, N. 1934. Dissolution of gold and silver in cyanide solutions. *Transactions of the AIME.* 112: 660–677.

Bodländer, G. 1896. Ueber abnorme gefrierpunktserniedrigungen. *Zeitschrift für Physikalische Chemie.* 21: 378–382.

Boonstra, B. 1943. Uber die losungsgeschwindigkeit von gold in kalium cyanidlosungen. Korros. *Metallschutz.* 19: 146–151.

Botz, M. 1999. *Overview of Cyanide Treatment Methods.* The Gold Institute, Washington, DC.

Botz, M.M., Mudder, T.I. 2000. Modeling of natural cyanide attenuation in tailings impoundments. *Minerals and Metallurgical Processing.* 17(4): 228–233.

Bucsh, O.R., Spottiswood, D.J., Lower, G.W. 1980. Ion-precipitate flotation of iron-cyanide complexes. *Water Pollution Control Federation.* 52(12): 2925–2930.

Campbell, J., Watson, B. 2003. Gravity leaching with the ConSep ACACIA reactor results from Anglogold unions reefs. In: *Proceedings of 8th Mil Operators Conference,* pp. 167–175. http://knelsongravity.xplorex.com/sites/knelsongravity/files/reports/report43s.pdf.

Carrillo-Pedroza, F.R., Soria-Aguilar, M.J. 2001. Destruction of cyanide by ozone in two gas-liquid contacting systems. *ejmp & ep (European Journal of Mineral Processing and Environmental Protection).* 1(1): 55–63.

Carrillo-Pedroza, F.R., Soria-Aguilar, M.J. 2001. Destruction of cyanide by ozone in two gas-liquid contacting systems. *The European Journal of Mineral Processing and Environmental Protection.* 1: 55–61.

Christy, S. 1896. The solution and precipitation of the cyanide of gold. *Transactions of the AIME.* 26: 735–772.

Cornejo, L.M., Spottiswood, D.J. 1984. Fundamental aspects of the gold cyanidation process: A review. *Mineral and Energy Resources (United States).* 27(2): 1–8.

Crowe, J.H.V. 1918. *General Smuts' Campaign in East Africa.* John Murray, London.

Dai, X., Jeffrey, M.I. 2006. The effect of sulfide minerals on the leaching of gold in aerated cyanide solutions. *Hydrometallurgy.* 82(3): 118–125.

Dash, R.B., Balomajumder, C., Kumar, A. 2008. Treatment of metal cyanide bearing wastewater by simultaneous adsorption and biodegradation (SAB). *Journal of Hazardous Materials.* 152(1): 387–396.

Davidson, R. 1974. The mechanism of gold adsorption on activated charcoal. *Journal of the Southern African Institute of Mining and Metallurgy.* 75: 67–76.

Davidson, R., Douglas, W., Tumilty, J. 1982. Aspects of laboratory and pilot-plant evaluation of CIP with relation to gold recovery. In: *CIM Bulletin.* Canadian Institute of Mining Metallurgy Petroleum, Calgary, AB, p. 73.

Davies, R.M., Mackenzie, W.M., Sole, C.K., Virnig, J.M. 1998. Recovery of Cu and CN- by SX from solutions produced in leaching of Cu/Au ores. *Copper Hydrometallurgy Forum.* 2: 16.

Davis, W.M. 1880. Depositing gold from its solutions. U.S. Patent. 227:963.

Deschênes, G., Lacasse, S., Fulton, M. 2003. Improvement of cyanidation practice at goldcorp Red Lake Mine. *Minerals Engineering*. 16: 503–509.

Deschênes, G., Lastra, R., Brown, J.R., Jin, S., May, O., Ghali, E. 2000. Effect of lead nitrate on cyanidation of gold ores: Progress on the study of the mechanisms. *Minerals Engineering*. 13: 1263–1279.

Durney, J.L. 1984. *Electroplating Engineering Handbook*, 4th edition. Van Nostrand Reinhold, New York.

Dzombac, D., Ghosh, R.A.-C. 2006. *Cyanide in Water and Soil: Chemistry, Risk and Management*. Taylor and Francis/CRC, Boca Raton, FL.

Eisele, J., Wroblewski, M., Elges, M., McCleland, G. 1986. Staged heap leaching and direct electrowinning. USBM. IC. 9059.

Eisler, R., Wiemeyer, S.N. 2004. Cyanide hazards to plants and animals from gold mining and related water issues. *Reviews of Environmental Contamination and Toxicology*. 183: 21–54.

Ellis, S., Senanayake, G. 2004. The effects of dissolved oxygen and cyanide dosage on gold extraction from a pyrrhotite-rich ore. *Hydrometallurgy*. 72: 39–50.

Elmore, C., Brison, R., Kenny, C. 1988. The kamyr cilo process. *Perth Gold 88*: 197.

Elsner, L. 1846. Beobachtungen über das verhalten regulinischer metalle in einer wässrigen lösung von cyankalium. *Advanced Synthesis and Catalysis*. 37: 441–446.

EPA: US, U.E. 1994. *Treatment of Cyanide Heap Leaches and Tailings*. Office of Solid Waste, Special Waste Branch, Washington, DC.

EPRI, C.E. 1997. *Membrane Technologies for Water and Wastewater Treatment*. Electric Power Research Institute, Palo Alto, CA.

Eugene, W.W.L., Mujumdar, A.S. 2009. Gold extraction and recovery processes. Minerals, Metals, and Materials Technology Centre, National University of Singapore.

Faraday, M. 1857. The Bakerian lecture: Experimental relations of gold (and other metals) to light. *Philosophical Transactions of the Royal Society of London*. 147:145–181.

Fast, J.L. 1988. Carbon-in-pulp pioneering at the Carlton mill. How CIP processing blossomed as a routine at golden cycle in the 1950s. *E&MJ-Engineering and Mining Journal*. 189(6): 56–57.

Finkelstein, N. 1972. The chemistry of the extraction of gold from its ores. In: Adamson, R.J. (Ed.), *Gold Metallurgy on the Witwatersrand*. Cape and Transvaal Printers Ltd, Cape Town, South Africa, pp. 284–351.

Fisher, J., Weimer, D. 1964. *Precious Metals Plating*. R. Draper Ltd, Teddington, UK.

Follis, R. 1992. Assessing activated carbon quality in hydrometallurgical circuits: Analysis and presentation of data. In: *Proceedings of the Randol Gold Forum '92*. Randol International, Golden, CO, pp. 469–476.

Gogate, P.R., Pandit, A.B. 2004. A review of imperative technologies for wastewater treatment I: Oxidation technologies at ambient conditions. *Advances in Environmental Research*. 8(3): 501–551.

Gray, S., Katsikaros, N., Fallon, P. 2000. Gold recovery from copper gold gravity concentrates using the inline leach reactor and weak base resin. In: *Processing of Copper Gold Ores, Proceedings of Oretest Copper Gold Symposium*, Perth, WA, pp. 67–80.

Green, M. 1913. The action of oxidisers in cyaniding. *Journal of the Chemical, Metallurgical and Mining Society*. 13: 355.

Griffiths, S., Knorre, H., Gos, S., Higgins, R. 1987. The detoxification of gold- mill tailings with hydrogen peroxide. *Journal of the South African Institute of Mining and Metallurgy*. 87(9): 279–283.

Gross, J., Scott, J.W. 1927. Precipitation of gold and silver from cyanide solution on charcoal. USGPO, Washington, DC.

Guo, H., Deschênes, G., Pratt, A., Fulton, M., Lastra, R. 2004. Leaching kinetics and mechanisms of surface reactions during cyanidation of gold in presence of pyrite and stibnite. In: Annual SME meeting 2004. Preprint 04-73. The Society of Mining, Metallurgy and Exploration, Inc., Littleton, CO.

Gurbuz, N., Ozdemir, I., Demir, S., Cetinkaya, B. 2004. Improved palladium catalysed coupling reactions of aryl halides using saturated N-heterocarbene ligands. *Journal of Molecular Catalysis. A: Chemical.* 209(1): 23–28.

Habashi, F. 1967. *Kinetics and Mechanism of Gold and Silver Dissolution in Cyanide Solution.* Montana College of Mineral Science and Technology, Butte, MT.

Habashi, F. 1987. One hundred years of cyanidation. *Canadian Institute of Mining, Metallurgy and Petroleum Bulletin.* 80(905): 108–114.

Hedley, N., Tabachnick, H. 1958. *Chemistry of Cyanidation.* American Cyanamid Company, Explosives and Mining Chemicals Department, New York.

Huiatt, J.L. 1984. Cyanide from mineral processing: Problems and research needs. In: *Conference on Cyanide and the Environment*, Tucson, AZ. Geotechnical Engineering Program. Colorado State University, Fon Collins, CO.

IAEA. 2002. Application of ion exchange processes for the treatment of radioactive waste and management of spent ion exchangers. IAEA, Vienna, Austria. ISBN: 92-0-112002-8.

Iordache, I., Nechita, M., Aelenei, N., Rosca, I., Apostolescu, G., Peptanariu, M. 2003. Sonochemical enhancement of cyanide ion degradation from wastewater in the presence of hydrogen peroxide. *Polish Journal of Environmental Studies.* 12(6): 735–737.

Janin, L. Jr., 1888. Cyanide of potassium as a lixiviate agent for silver ores and mineral. *Engineering and Mining Journal.* 46: 548–549.

Janin, L. Jr., 1892. The cyanide process. Mineral Industry. 1: 239–272.

Jara, J., Harris, R. 1994. A new device to enhance oxygen dispersion in gold cyanidation. In: *Proceedings, Annual Meeting of the Canadian Mineral Processors*, Paper 94. Canadian Institute of Mining, Metallurgy and Petroleum (CIM), Montreal, QC.

Jin, S., May, O., Ghali, E., Deschênes, G. 1998. Investigation on the mechanisms of the catalytical effect of lead salts on gold dissolution in cyanide solution. In: Xianwan, Y., Qiyuan, C., Aiping, H. (Eds), *Third International Conference on Hydrometallurgy ICHM '98.* International Academic Publishers, Beijing, China, pp. 666–679.

Johnson, W. 1894. Method of abstracting gold and silver from their solutions in potassium cyanides. US Patent. 522:260.

Julian, H.F., Smart, E. 1921. *Cyaniding Gold and Silver Ores.* C. Griffin and Co., London.

Kawatra, S.K. 2002. *Froth Flotation: Fundamental Principles.* Research, Michigan Technical University, Houghton, MI.

Kim, Y.J., Qureshi, T.I., Min, K.S. 2003. Application of advanced oxidation processes for the treatment of cyanide containing effluent. *Environmental Technology.* 24: 1269–1276.

Kondos, P.D., Deschenes, G., Morrison, R.M. 1995. Process optimization studies in gold cyanidation. *Hydrometallurgy.* 39: 235–250.

La Brooy, S., Bax, A., Muir, D., Hosking, J., Hughes, H., Parentich, A. 1986. Fouling of activated carbon by circuit organics. In: *Gold 100: Proceedings of the International Conference on Gold*, SAIMM, Johannesburg, South Africa, pp. 123–132.

La Brooy, S., Komosa, T., Muir, D. 1991. Selective leaching of gold from copper-gold ores using ammonia-cyanide mixtures. In: *Proceedings of 5th AusIMM Extractive Metallurgy Conference*, Perth, WA.

Latkowska, B., Figa, J. 2007. Cyanide removal from industrial wastewaters. *Polish Journal of Environmental Study.* 16(2A): 748–752.

Lee, K.C.-K. 2005. Cyanide regeneration from thiocyanate with the use of anion exchange resins. Chemical Engineering, University of New South Wales, Sydney, Australia.

Lemos, F., Gonzoga, S., Dutra, J. 2006. Copper electrowinning from gold plant waste streams. *Minerals Engineering.* 19(5): 388–398.

Ling, P., Papangelakis, V.G., Argyropoulos, S.A., Kondos P. D. 1996. An improved rate equation for cyanidation of a gold ore. *Canadian Metallurgical Quarterly.* 35: 225–234.

Liu, G.O., Yen, W.T. 1995. Dissolution kinetics and carbon adsorption for the cyanidation of gold ores in oxygen-enriched slurry. *Canadian Institute of Mining, Metallurgy and Petroleum Bulletin.* 88(986): 42–48.

Logsdon, M.J., Hagelstein, K., Mudder, T.I. 1999. *The Management of Cyanide in Gold Extraction.* International Council on Metals and the Environment, Ottawa, ON.

Longley, R.J., McCallum, A., Katsikaros, N. 2003. Intensive cyanidation: Onsite application of the InLine Leach Reactor to gravity gold concentrates. *Minerals Engineering.* 16(5): 411–419.

Loroesch, J., Knorre, H., Merz, F., Gos, S., Marais, H.J. 1988. The Degussa PAL system-a future technology in cyanidation. *Perth Gold.* 88: 202.

MacArthur, J.S. 1905. Gold extraction by cyanide: A retrospect. *Journal of the Society of Chemical Industry.* 24: 311–315.

Maclaurin, R.C. 1893. XLVIII—The dissolution of gold in a solution of potassium cyanide. *Journal of the Chemical Society, Transactions.* 63: 724–738.

Marsden, J., House, C.I. 1992. *The Chemistry of Gold Extraction.* Ellis Horwood, New York.

Marsden, J.O., House, C.I. 2006. *The Chemistry of Gold Extraction*, 2nd Edition. The Society of Mining, Metallurgy, and Exploration Inc. (SME), Littleton, CO.

McArthur, J., Forrest, R., Forrest, W. 1887. Process obtaining gold and silver from ores. British Patent 14174.

McLaughlin, J.D., Quinn, P., Agar, G.E., Cloutier, J.Y., Dub, G., Leclerc, A. 1999. Oxygen mass transfer rate measurements under different hydrodynamic regimes. *Industrie Minerale Mines et Carrieres les Techniques.* 76(3–4): 121–126.

McMullen, J., Thompson, R. 1989. Practical use of oxygen for gold leaching in Canada. In: *Randol Gold and Silver Recovery Innovations: Phase IV Workshop*, Sacramento, CA, pp. 99–100.

Miller, J.D. 2003. Treatment of cyanide solutions and slurries using air-sparged hydrocyclone (Ash) technology. The Office of Industrial Technologies, Energy Efficiency and Renewable Energy U.S. Department of Energy, Salt Lake, UT.

National Risk Management Research Laboratory (U.S.). 2000. *Managing Cyanide in Metal Finishing.* U.S. Environmental Protection Agency, Office of Research and Development, National Risk Management Research Laboratory, Technology Transfer and Support Division, Cincinnati, OH.

Nicol, H. 1979. A modern study of the kinetics and mechanism of the cementation of gold. *Journal of the Southern African Institute of Mining and Metallurgy.* 79: 191–198.

Oelsner, K., Dornid, D., Uhlemann, R. 2001. Degradation of complex cyanides. Bioservice Waldenburg. Saxon State Ministry of the Environment and Agriculture.

Osei-Agyemang, E., Paul, J.F., Lucas, R., Foucaud, S., Cristol, S. 2015. Stability, equilibrium morphology and hydration of zrc (111) and (110) surfaces with H_2O: A combined periodic dft and atomistic thermodynamic study. *Physical Chemistry Chemical Physics.* 17(33): 21401–21413.

Osmonics, I. 1997. *Pure Water Handbook.* Osmonics Inc., Minnetonka, MN.

Oxomax. 2005. *Advanced Oxidation Process.* http://www.ozomax.com/pdf/Article_on_AOP_Pheno_destruct.pdf.

Pargaa, J., Shuklab, S., Carrillo-Pedrozac, F. 2003. Destruction of cyanide waste solutions using chlorine dioxide, ozone and titania sol. *Waste Management.* 23: 183–191.

Palmer, G.R. 1986. Ion-exchange research in precious metals recovery. In: *Proceedings of Precious Metals Recovery from Low-Grade Resources,* U.S. Bureau of Mines, Washington, DC, pp. 2–9.

Park, J. 1898. Notes on the action of cyanogen on gold. *Transactions of the American Institute of Mining and Metallurgical Engineers.* 6: 120.

Revy, T., Watson, S., Hoecker, W. 1991. Oxygen assisted cyanidation of gold in Australia. In: *Randol Gold Forum '91,* Cairns, Australia. Randol International, Golden, CO, pp. 317–324.

Riani, J.C., Leao, V.A., Silva, C.A.D., Silva, A.M., Bertolino, S.M., Lukey, G.C. 2007. The elution of metal cyanocomplexes from poly acrylic and polystere based ion exchange resins using nitrate and thiocyanate eluants. *Brazlian Journal of Chemical Engineering.* 24(3): 421–431.

Riveros, P., Cooper, W. 1987. Ion exchange of gold and silver from cyanide solutions. In: *Proceedings of the International Symposium on Gold Metallurgy,* Pergamon Press, New York, p. 379.

Sandstrom, A., Shchukarev, A., Paul, J. 2005. XPS characterization of chalcopyrite chemically and bioleached at high and low redox potential. *Minerals Engineering.* 18(5): 505–515.

Sceresini, B. 1997. The Filblast cyanidation process—A maturing technology. In: *Randol Gold Forum '97,* Monterey, CA. Randol International, Golden, CA, pp. 173–179.

Scott, S.J. 1984. An overview of cyanide treatment methods for gold mill effluents. In: *Conference on Cyanide and the Environment,* Tucson, AZ. Geotechnical Engineering Program, Colorado State University, Fon Collins, CO.

Shipman, A. 1994. Laboratory methods for the testing of activated carbon for use in carbon-in-pulp plants for the recovery of gold. In: *Mint– Communication MC1,* Mint– Corporation, Randburg, South Africa.

Simeonova, F.P., Fishbein, L., World Health Organization. 2004. Hydrogen cyanide and cyanides: Human health aspects. World Health Organization, Geneva, Switzerland.

Simovic, L., Snodgrass, W.J. 1999. Techniques for evaluating a mathematical model for the natural degradation of cyanide from gold mill effluents. http://pdf.library.laurentian.ca/medb/conf/sudbury99/newtech/ntop14.pdf.

Skey, W. 1897. A note on cyanide process. *Engineering and Mining Journal.* 63.

Sobral, G.S., Dutra, A.B., Rosenberg, B., Lemos, F. 2002. Cyanide electrolytic recycling. Green Processing Congress. Centro De Tecnologia Mineral Centro De Tecnologia Mineral Coordenação De Metalurgia Extrativa, Kern-Australia, Kew East, VIC, pp. 147–150.

Sorokin, D.Y., Tourova, T.P., Lysenko, A.M., Kuenen, J.G. 2001. Microbial thiocyanate utilization under highly alkaline conditions. *Applied Biochemistry Microbiology.* 67: 528–538.

Stephens, T. 1988. The use of pure oxygen in the leaching process in South African gold mines. In: *Perth International Gold Conference.* Randol International, Golden, CO, pp. 191–196.

Terry, I.M., Botz, M.M., Smidth, A. 2001. *Chemistry and Treatment of Cyanidation Wastes,* 2nd Edition. Mining Journal Books Limited, London.

The United States Environmental Protection Agency. 2017. Ground water and drinking water. https://www.epa.gov/ground-water-and-drinking-water/national-primary-drinking-water-regulations.

USEPA, U.S. 2000. Managing cyanide in metal finishing. Retrieved December 01, 2010. http://Printfu.Org/Read/CapsuleReportManagingCyanideInMetalFinishing 9e16.Html?F=1qeypurpn6wihSupogum6wnh6_Q5tjp1mqqxm7e49jpjbln3dfm 3bmklxizLP2dKF1d2WstnxtySr9fizJvztrwmkdnopeqln3ilqhop5adqz jp6dvz3–5vtns2oG36mnn5boh6vlo6apojfv2edfsjsj39znonze.

Volesky, B., Naja, G. 2005. Biosorption application strategies. In: *16th International Biotechnology Symposium.* Compress Co., Cape Town, South Africa.

Wadsworth, M.E., Zhu, X., Thompson, J.S., Pereira, C.J. 2000. Gold dissolution and activation in cyanide solution: Kinetics and mechanism. *Hydrometallurgy.* 57: 1–11.

Wang, L.K., Hung, Y.-T., Howard, H.L., Constantine, Y. 2006. *Handbook of Industrial and Hazardous Waste Treatment.* Marcel D–ker, Inc., New York.

Wilkinson, P. 1986. Understanding gold plating. *Gold Bulletin.* 19: 75–81.

Woollacott, L., Erasmus, C. 1992. The distribution of gold on loaded carbon. *Journal of the South African Institute of Mining and Metallurgy.* 92(7): 177–182.

Young, G., Douglas, W., Hampshire, M. 1984. Carbon in pulp process for recovering gold from acid plant calcines at president brand. *Mining Engineering.* 36: 257–264.

Young, C.A., Jordan, T.S. 1995. Cyanide remediation: Current and past technologies. In: *Conference on Hazardous Waste Research,* Manhattan, KS. Department of Metallurgical Engineering, Montana Tech, Butte, MT.

Zadra, J. 1950. A process for the recovery of gold from activated carbon by leaching and electrolysis. U.S. Department of the Interior, Bureau of Mines, Washington, DC.

Zadra, J., Engel, A.L., Heinen, H.J. 1952. Process for recovering gold and silver from activated carbon by leaching and electrolysis. U.S. Department of the Interior, Bureau of Mines, Washington, DC.

4

Thiosulfate Leaching of Gold

Sadia Ilyas[*], Aqsa Zia[*], and Jae-chun Lee[†]

4.1 Introduction

Thiosulfate salts have been used in industry for more than two and a half centuries. They have the chemical symbol $S_2O_3^{2-}$ and can be produced from elemental sulphur and sulphite at elevated temperature:

$$S + SO_3^{2-} = S_2O_3^{2-} \tag{4.1}$$

Under alkaline conditions, thiosulfate can be produced as a product of the reaction between sulphur or sulphide and hydroxide as described by Equations 4.2 and 4.3 (Shi> et al., 1965).

$$S_8 + 8NaOH = 2Na_2S_2O_3 + 4NaHS + 2H_2O \tag{4.2}$$

$$2(NH_4)_2S_5 + 6NH_4OH + 6O_2 = 5(NH_4)_2S_2O_3 + 3H_2O \tag{4.3}$$

Thiosulfate is tetrahedral in shape and can be derived by replacing one oxygen atom with a sulphur atom in a sulphate anion (Figure 4.1). The sulphur-to-sulphur (S–S) distance indicates a single bond, implying that the sulphur bears a significant negative charge and the S–O interactions have more than double the bond character. The thiosulfate anion is meta-stable; a natural donor of sulphur that disproportionately forms sulphite and sulphur or an active sulphur species. It is sometimes regenerated from tri, tetra, or penta thionates in an alkaline aqueous ammonia solution, as shown in Equations 4.4–4.8 (Naito et al., 1970; Aylmore and Muir, 2001; Zhang and Dreisinger, 2002).

[*] Mineral and Material Chemistry Lab, Department of Chemistry, University of Agriculture Faisalabad, Pakistan.
[†] Minerals Resources Research Division, Korea Institute of Geoscience and Mineral Resources, Daejeon, South Korea.

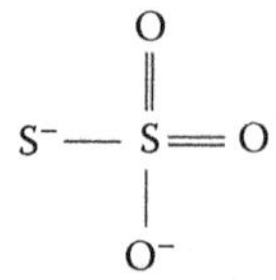

$$\underset{\underset{O^-}{|}}{\overset{\overset{\displaystyle O}{\|}}{S^- {-}\ S}}{=}O$$

FIGURE 4.1
Chemical structure of the thiosulfate ion.

$$4S_4O_6^{2-} + 6OH^- = 2S_3O_6^{2-} + 5S_2O_3^{2-} + 3H_2O \qquad (4.4)$$

$$2S_4O_6^{2-} = S_3O_6^{2-} + S_5O_6^{2-} \qquad (4.5)$$

$$2S_5O_6^{2-} + 6OH^- = 5S_2O_3^{2-} + 3H_2O \qquad (4.6)$$

$$S_3O_6^{2-} + 2OH^- = SO_4^{2-} + S_2O_3^{2-} + H_2O \qquad (4.7)$$

$$S_3O_6^{2-} + NH_3 + OH^- = SO_3NH_2^- + S_2O_3^{2-} + H_2O \qquad (4.8)$$

The compatibility of an environmentally benign thiosulfate ligand when combined with gold, along with the achievable fast kinetics similar to cyanidation, are some of the basic characteristics that present thiosulfate as the prime alternative candidate in gold metallurgy. Especially in the case of carbonaceous ore cyanidation, the *in* situ adsorption of anionic gold cyanide complex onto the carbonaceous matter can be avoided using thiosulfate as lixiviant. The affinity order of gold adsorption on carbon surface is: $SCN^- > SC(NH_2)_2 > CN^- S_2O_3^{2-}$. Although thiosulfate exhibits the anionic species, possibly $Au(S_2O_3)^-$ and $Au(S_2O_3)_2^{3-}$, it shows less affinity for the reductive adsorption of the thiosulfate complex, as shown with the adsorption of cyanide and chloride complexes on carbon surface, and leads to treatment of carbonaceous ores in an environmentally benign manner.

4.2 History of Gold-Thiosulfate Leaching

The recovery of gold using thiosulfate was first proposed early in the 1900s (White, 1900). In a process known as the Von Patera process, gold and silver ores were subjected to a chlorination roasting followed by leaching in a thiosulfate solution. In South America, thiosulfate leaching of the roasted mass of silver-rich sulphide ores was in practice around the time of World War II (Flett et al., 1983). A similar treatment was also carried out at the Colorado Mine at Sonora in Mexico (Von Michaelis, 1987). However, it was not until the late 1970s that an application to recover precious metals from copper-bearing metal sulphides concentrates and pressure leach residues was developed

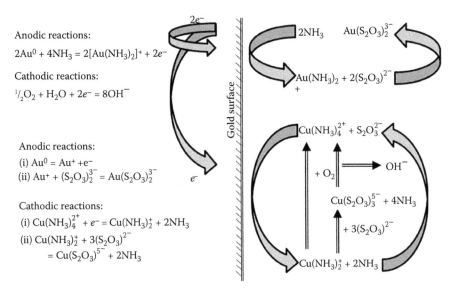

FIGURE 4.2
Electrochemical mechanism for golf leaching in a thiosulfate solution.

employing ammonium thiosulfate and patented by Berezowsky and Kerry (Brent Hiskey and Atluri, 1988). During this period, studies were also carried out in the former Soviet Union (Ter-Arakeyan et al., 1984) and demonstrated the presence of copper ions in solution could fasten the gold dissolution. Berezowsky and Sefton (1979) revived interest in thiosulfate leaching by developing an atmospheric ammoniacal thiosulfate leach process to recover gold and silver from residues of the ammonia oxidation leaching of sulphide copper concentrates. Tozawa et al. (1981) and Jiang et al. (1993) focused on the kinetics and mechanism of gold dissolution by thiosulfate, as shown in Figure 4.2. Heap leaching of carbonaceous preg-robbing ores using thiosulfate was established by Newmont Gold Company. Unlike gold cyanide, gold thiosulfate does not adsorb on carbonaceous material and much higher gold recoveries were achieved.

4.3 Influential Factors for Extracting Gold in a Thiosulfate Solution

The chemistry of the gold extraction in thiosulfate medium is complex, and requires an oxidizing environment to keep reactions under control. The redox couple reaction of Cu^+ to Cu^{2+} and vice-versa with its self-catalytic bιaviour is used for this purpose (Umetsu and Tosawa, 1972). Thiosulfate decomposition in the presence of acid is commonly prevented by using an ammoniacal medium, under which conditions copper amine complexes catalyse the leaching kinetics (Aylmore and Muir, 2001). The formation of gold complexes in thiosulfate solution in the presence of ammonia and the redox couple of Cu^+ and Cu^{2+} can be represented by the Eh-pH diagram as shown in Figure 4.3a and b.

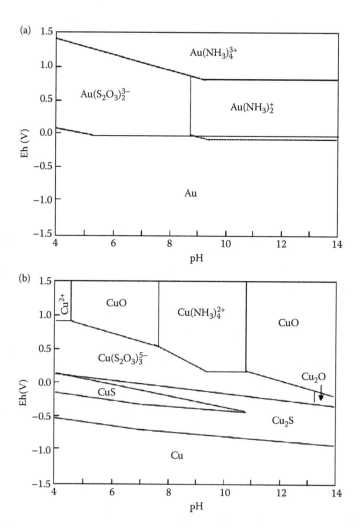

FIGURE 4.3

(a) Eh-pH diagram of gold complexes in thiosulfate media in the presence of ammonia and (b) Eh-pH diagram of gold complexes in thiosulfate media in the presence of cuperous/cupperric redox couple.

It can be seen that the formation of two complexes $Au(S_2O_3)^-$ and $Au(S_2O_3)_2^{3-}$ is possible, in which the formation of the most stable complex takes place as (Johnson and Davis, 1973):

$$4Au + 8S_2O_3^{2-} + O_2 + 2H_2O = 4[Au(S_2O_3)_2]^{3-} + 4OH^- \qquad (4.9)$$

4.3.1 Effect of Ammonia

Gold complexation with thiosulfate following the previous reaction exhibits slow kinetics due to decomposition of thiosulfate to sulphur passivates onto gold surfaces, which hinders the leaching of gold (Pedraza et al., 1988; Jiang et al., 1993; Chen et al., 1996). Alkaline solutions must be used to prevent thiosulfate dissolution; therefore, ammonia is used which also prevents the surface passivation and enhances the gold complexation as follows (Jiang et al., 1993; Chen et al., 1996):

$$Au(NH_3)^{2+} + 2S_2O_3^{2-} = [Au(S_2O_3)_2]^{3-} + 2NH_3 \qquad (4.10)$$

The presence of ammonia preferentially hinders the dissolution of the most common gangue minerals (silica, silicates, and carbonates) and other base metal impurities, like iron in the gold bearing ores (Abbruzzese et al., 1995). However, since the above reaction is thermodynamically feasible in an ammonia solution, the favourable kinetic is essentially not achievable at ambient temperature (Meng and Han, 1993). It requires a higher temperature (>80°C) although this is less attractive in a process economy and has a higher decomposition rate of ammonia above 60°C (Tozawa et al., 1981). The slower kinetics at ambient temperatures can be encountered by the catalytic action of copper ions in a gold-thiosulfate leaching (Tyurin and Kakowski, 1960), limiting the role of ammonia mainly to stabilize the Cu^{2+}.

4.3.2 Effect of Copper

The catalytic action of Cu^{2+} to fasten the gold leaching in thiosulfate solution was first reported by Tyurin and Kakowsky (1960), and significantly yields a leaching rate 18–20 folds higher than without employing Cu^{2+} (Ter-Arakeyan et al., 1984). The reported increase in dissolution of gold in copper thiosulfate solutions containing ammonia has been attributed to the formation of $Cu(NH_3)_4^{2+}$ complexes. The amine complex, $Cu(NH_3)_4^{2+}$, forms at an even lower temperature than the ammonia decomposition temperature (<60°C). In such case, the leaching reaction drives as the reaction below:

$$Au + Cu(NH_3)_4^{2+} = Au(NH_3)_2^+ + Cu(NH_3)_2^+ \qquad (4.11)$$

The role of Cu^{2+} ions in the oxidation of gold to aurous ion (Au^+) can be understood as:

$$Au + 5S_2O_3^{2-} + Cu(NH_3)_4^{2+} = Au(S_2O_3)_2^{3-} + 4NH_3 + Cu(S_2O_3)_3^{5-} \quad (4.12)$$

Nevertheless, in the presence of Cu^{2+} the gold-thiosulfate-ammonia leaching is an electro-chemical reaction in which the conversion of $Cu(NH_3)_4^{2+}$ to $Cu(NH_3)_2^+$ supports the product formation of $[Au(S_2O_3)_2]^{3-}$. The electrochemical reactions can be given as:

$$Au + 2S_2O_3^{2-} = [Au(S_2O_3)_2]^{3-} + e^- \quad (4.13)$$

$$Cu(NH_3)_4^{2+} + 3S_2O_3^{2-} + e^- = [Cu(S_2O_3)_3]^{5-} + 4NH_3 \quad (4.14)$$

$$2Cu(NH_3)_4^{2+} + 8S_2O_3^{2-} = 2[Cu(S_2O_3)_3]^{5-} + S_4O_6^{2-} + 8NH_3 \quad (4.15)$$

The concentration of Cu^{2+} in the leaching system is a vital factor in thiosulfate stability and the management of lixiviant concentration. The reduction of Cu^{2+} in the presence of thiosulfate ions is extremely fast in a pure system; however, it slows in ammoniacal solution and depends on the ammonia concentration.

4.3.3 Effect of Oxygen

The completion of redox equilibrium between the cuprous-cupric couple as described in the previously discussed electro-chemical reactions would require the conversion of Cu^+ ion to Cu^{2+} to drive further leaching of gold. It is fulfilled by the oxygen supply to the system to proceed the reaction as follows (Abbruzzese et al., 1995):

$$2Cu(S_2O_3)_3^{5-} + 8NH_3 + \frac{1}{2}O_2 + H_2O = 2Cu(NH_3)_4^{2+} + 2OH^- + 6S_2O_3 \quad (4.16)$$

The amount of dissolved oxygen in thiosulfate solution of ammoniacal medium directly affects the rapid oxidation of Cu^+ to Cu^{2+}, along with a little oxidation of thiosulfate forming sulphate and trithionate (Byerley et al., 1973, 1975). Notably, the oxidation of thiosulfate by oxygen in aqueous media under ambient temperature and pressure is known to be extremely slow and only prevails in the presence of Cu^{2+} ions and ammonia (Naito et al., 1970). In contrast to the presence of oxygen, Cu^{2+} ions initially oxidize thiosulfate ions to tetrathionate ions, which subsequently undergo a disproportionation reaction to yield the trithionate and thiosulfate ions. The decomposition of thiosulfate yields black precipitates of copper sulphides at low potentials where oxidants are deficient, in stagnant solutions, or in high copper containing solutions. Hence, the precipitation of copper sulphides is related to the availability of oxygen in the system. A limited solubility of oxygen in an aqueous medium and the slow reduction at the gold surface makes the use of oxygen without the copper catalytic reaction very slow, resulting in low gold dissolution.

4.4 Species Present in Thiosulfate Leach Liquors

4.4.1 Metal Complexes

Thiosulfate is a divalent soft ligand and tends to form stable complexes with low-spin d^{10} (Au^+, Ag^+, Cu^+, Hg^{2+}) and d^8 (Au^{3+}, Pt^{2+}, Pd^{2+}) metal ions (Livingstone, 1965; Wilkinson and Gillard, 1987). Mostly, the thiosulfate ion acts as a unidentate ligand via the terminal sulphur atom, establishing strong σ bonds with a metal ion, which are stabilized by $p\pi$–$d\pi$ back-bonding (Figure 4.4). Thiosulfate ligands may also act in a bridging role via the terminal sulphur atom or as a bidentate ligand through a sulphur and an oxygen atom, usually resulting in an insoluble complex (Figure 4.5) (Ryabchikov, 1943; Livingstone, 1965; Gmelin, 1973; Zhao and Wu, 1997).

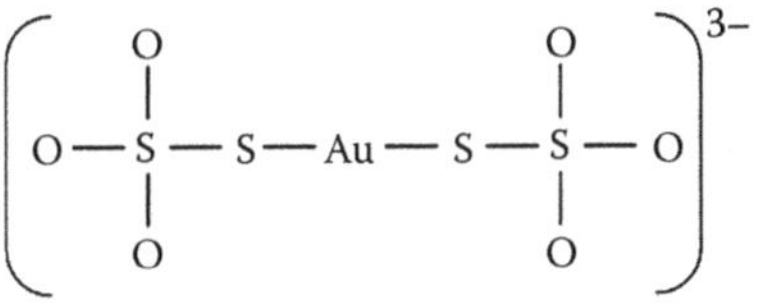

FIGURE 4.4
Structure of gold thiosulfate unidentate complex.

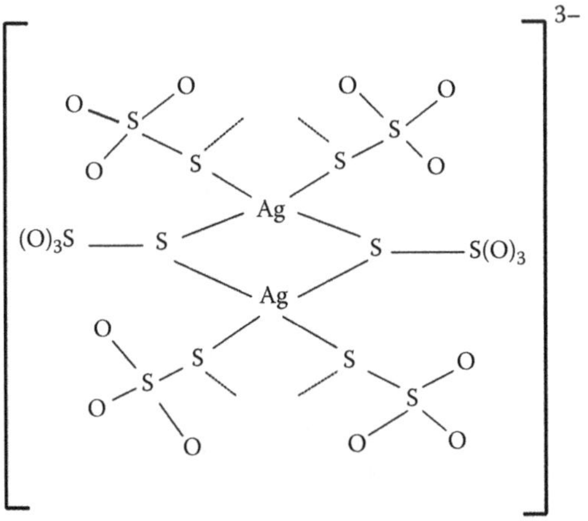

FIGURE 4.5
Bridging complex of thiosulfate with silver.

In a thiosulfate leach liquor, the formation of gold and silver thiosulfate complexes proceeds via the catalytic oxidation of the zero-valent metal by a suitable soluble metal complex, which is typically the $[Cu(NH_3)_4]^{2+}$ complex, acting as the primary oxidant (Tozawa et al., 1981; Jiang et al., 1993a; Li et al., 1995). In ammoniacal thiosulfate liquors, metal ions can form a range of complexes. In the case of copper, Cu^+ is commonly reported as $[Cu(S_2O_3)_3]^{5-}$, yet at concentrations of thiosulfate below 0.05 M, the primary complex is expected to be $[Cu(S_2O_3)_2]^{3-}$ (Naito et al., 1970; Zipperian et al., 1988). Whereas Cu^{2+} dominantly exists as $[Cu(NH_3)_4]^{2+}$, although some other amine species may form (Byerley et al., 1973, 1975). It appears to be the presence of the maximum solubility of copper, in an approximation of one gram of soluble copper with 1% $(NH_4)_2S_2O_3$ (w/w) (Johnson and Bhappu, 1969). In unfavourable conditions, the precipitation of $Cu_2S_2O_3$, or mixed salts of cuprous-ammonium thiosulfate, may occur (Flett et al., 1983; Chen et al., 1996). Soluble thiosulfate complexes are also known for a number of heavy metals, with their stepwise stability constants and coordination numbers illustrated in Figure 4.6 (Vasil'ev et al., 1953; Novakovskii and Ryazantseva, 1955; Gmelin, 1965; Livingstone, 1965; Gmelin, 1969, 1972, 1973; Wilkinson and Gillard, 1987; Vlassopoulos and Wood, 1990; Tykodi, 1990; Benedetti and Boulëgue, 1991; Williamson and Rimstidt, 1993; Hubin and Vereecken, 1994).

Apart from the formation of anionic thiosulfate complexes, some metal cations are expected to form ammine complexes, as detailed in Figure 4.7 (Smith and Martell, 1976). Several copper and palladium complexes bearing both ammine and thiosulfate ligands are also known, although their

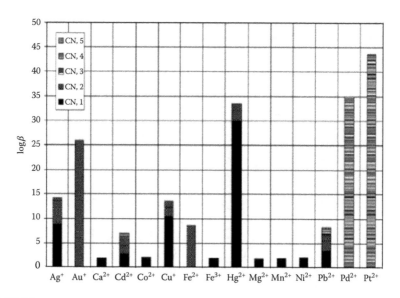

FIGURE 4.6

Stability constant and coordination number of thiosulfate complexes of metal ions.

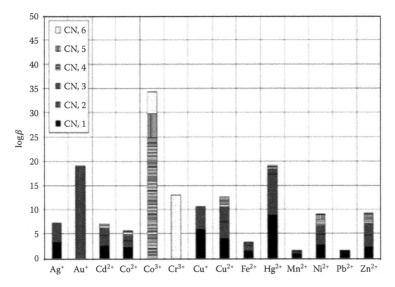

FIGURE 4.7
Stability constant and coordination number of amine complexes of metal ions.

stability constants and solubility remain unknown (Wilkinson and Gillard, 1987). Comparison of the stability constants in Figures 4.6 and 4.7 reveals that many important metals will form thiosulfate complexes in preference to their corresponding ammine complexes. Thiosulfate complexes are expected to predominate for Au^+, Ag^+, Fe^{2+}, Hg^{2+}, and Pb^{2+}, whereas the metal ions Cu^+ and Cd^{2+} are an equilibrium mixture of thiosulfate and ammine complexes. The remaining soluble metal ions occur primarily as ammine complexes. The ligands of the aurothiosulfate complex are believed to be quite labile, as near-stoichiometric quantities of cyanide added to a thiosulfate liquor were found to rapidly form the corresponding aurocyanide complex (Lulham and Lindsay, 1991; Marchbank et al., 1996). This is especially significant when the rapid reaction between cyanide and thiosulfate ions is taken into account. The complexes $[Pd(S_2O_3)_4]^{6-}$ and $[Pt(S_2O_3)_4]^{6-}$ have aqueous solubility above 10 ppb at pH 7 and 25°C, and quite low oxidation potentials (E^0) of −0.116 and −0.170 mV, respectively (Mountain and Wood, 1988; Plimer and Williams, 1988). However, they are not thermodynamically stable and slowly decompose into insoluble S-bridged oligomers similar to the silver complex in Figure 4.5 (Anthony and Williams, 1994). Metals in higher oxidation states such as Au^{3+} and Fe^{3+} are readily reduced by thiosulfate ions, and hence are not significant in leach liquors. Other anions from the leach liquor or the mineral matrix, such as chloride, hydroxide, or sulphate, may also participate in metal ion solvation. Stable, soluble complexes bearing a mixture of ligands may be present, similar to the copper salts $[Cu(CN)_x(NH_3)_y]^{(1-x)}$ found in ammoniacal cyanide leach liquors (Muir et al., 1993). Both mercury and silver tend to form nearly insoluble sulphides, although complexation in excess thiosulfate tends to minimize this precipitation (Bean, 1997). Many iron minerals, such as pyrite and haematite, and most TiO_2 and SiO_2 ores catalyse the oxidative degradation of thiosulfate ions into tetrathionate (Benedetti and Boulëgue, 1991; Xu and Schoonen, 1995). However, the side reactions and decomposition processes of many metal thiosulfate complexes are not well characterized. So, it is essential to observe natural degradation of thiosulfate via ubiquitous O_2, H_3O^+, trace Fe^{3+}, and other oxidants prior to mineral leaching. One important example is Cu^{2+}, which also contributes to the essential gold oxidation step.

4.4.2 Sulphur-Oxygen Anions

The sulphoxy anions initially present in an ammoniacal thiosulfate leach liquor are thiosulfate, sulphate, and sulphide from mineral sources. However, thiosulfate is metastable, which means it may be readily oxidized or reduced according to the initial solution potential. Depending on the aqueous environment, thiosulfate can break down into sulphite, sulphate, trithionate, tetrathionate, sulphide, polythionates ($S_xO_y^{2-}$), and/or polysulfides $x(S_x^{2-})$. An important factor in thiosulfate stability is the pH of the solution, since thiosulfate rapidly decomposes in acidic media (Li et al., 1995). Certain metal ions and reagents also cause the breakdown of thiosulfate, as shown in Equations 4.17–4.23 (Tykodi, 1990; Williamson and Rimstidt, 1993; Abbruzzese et al., 1995; Xu and Schoonen, 1995; Briones and Lapidus, 1998).

$$4S_2O_3^{2-} + O_2 + 4H^+ = 2S_4O_6^{2-} + 2H_2O \qquad (4.17)$$

$$S_2O_3^{2-} + 2H^+ = S^0 + SO_2 + 2H_2O \qquad (4.18)$$

$$4S_2O_3^{2-} + O_2 + 2H_2O = 2S_4O_6^{2-} + 4OH^- \qquad (4.19)$$

$$S_2O_3^{2-} + CN^- + 0.5O_2 = SCN^- + SO_4^{2-} \qquad (4.20)$$

$$2[Fe(S_2O_3)]^+ = 2Fe^{2+} + S_4O_6^{2-} \qquad (4.21)$$

$$S_2O_3^{2-} + Cu^{2+} + 2OH^- = SO_4^{2-} + H_2O + CuS \qquad (4.22)$$

$$2Cu^{2+} + 2S_2O_3^{2-} = 2Cu^+ + S_4O_6^{2-} \qquad (4.23)$$

Notably, the anions trithionate ($S_3O_6^{2-}$) and tetrathionate ($S_4O_6^{2-}$), which are not known to have any lixiviating activity (Aylmore, 2001), can interfere with resin-based recovery methods by displacing metal complexes from ion-exchange sites (Fleming, 1998; O'Malley, 2001). In addition to the previous reactions, thiosulfate is also consumed by peroxides, phosphines, polysulfides, permanganates, chromates, halogens (chlorine, bromine and iodine), and their oxyanions. In addition, certain species of fungi, micro fauna, and microflora can digest thiosulfate ions (Xu and Schoonen, 1995). The degradation of thiosulfate ions may be caused, or catalysed, by the presence of certain metal ions. The Fe^{3+} ion accelerates the decomposition of thiosulfate by intramolecular electron transfer. The deep purple $[Fe(S_2O_3)]^+$ complex is formed, and decomposition occurs via reduction of the metal and concomitant oxidation and dimerization of the ligand to form the tetrathionate ion (Uri, 1947; Perez and Galaviz, 1987; Williamson and Rimstidt, 1993). Similarly, the salts of arsenic, antimony, and tin catalyse the formation of pentathionate from thiosulfates, while metallic copper, zinc, and aluminium result in the formation of sulphides (Xu and Schoonen, 1995; Bean, 1997).

4.5 Recovery of Gold from Thiosulfate Leach Liquors

4.5.1 Precipitation

Reductive precipitation of gold from the pregnant leach liquor using inorganic zinc metal (the Merrill–Crowe process) or organic acid (prominently the oxalic acid) is a common process in gold recovery; however, it is not very effective in the thiosulfate-ammonia system. The metal precipitants often have a deleterious effect on thiosulfate ions, producing unwanted cations, and thus complicating the recycling process of lixiviant. The contamination of solid product is often a result of either undissolved (excess) precipitant or coprecipitation with other metal ions, necessitating further purification. Using copper is a reasonable choice, as the gold-depleted copper solution can be directly recycled to the leaching stage. Precipitation by the addition of sulphide salts or chemical reduction with sodium borohydride, hydrogen, or sulphur dioxide has also been investigated (Johnson and Bhappu, 1969; Deschenes and Ritcey, 1990; Awadalla and Ritcey, 1991). These techniques are not highly favoured, as they are less selective and tend to precipitate most metals from solution as well as hindering recycling of the leach liquor. The electro-reduction of aurothiosulfate ions to deposit on the cathode is especially problematic in the presence of a great excess of unwanted cations of copper, which are both deposited on the cathode product. This results in a devalued product requiring further purification. Side reactions involving the oxidation or reduction of thiosulfate may also interfere (Aylmore, 2001). This lowers the efficiency of electrowinning by increasing the energy input required to recover the desired metals from the solution, making it an unviable option for recovering the gold.

4.5.2 Carbon Adsorption

The affinity of aurothiosulfate complex for carbon appears to be in contention with the aurocyanide complex. The affinity order of gold adsorption on carbon was found to be: $SCN^- > SC(NH_2)_2 > CN^-S_2O_3^{2-}$. A 0.1 mol/L solution of the aurothiosulfate complex in KOH (10^{-4} mol/L) was found to have little or no affinity for carbon at 25°C (Gallagher et al., 1989). Conversely, in the studies carried out by Abbruzzese et al. (1995) and Meggiolaro et al. (2000), a recovery of 95% gold after 6 h contact at 25°C could be achieved from the gold-bearing (15.8 mg/L Au) leach liquor comprising: 2.0 mol/L $Na_2S_2O_3$ + 4.0 mol/L NH_3. This discrepancy may be due to a loss of sensitivity to small changes in the surprisingly high gold concentration in the latter converts the aurothiosulfate complex to the more stable aurocyanide and recovers that on a resin (Lulham and Lindsay, 1991; Marchbank et al., 1996).

4.5.3 Solvent Extraction

There have been a number of studies in which gold has been extracted from ammoniacal thiosulfate solutions by solvent extraction using a number of potential organic extractants diluted in various hydrocarbons (like benzene, kerosene, and octanols). In several cases, the presence of ammonia in gold solvent extraction has been found to improve the extraction efficiency (Zhao and Wu, 1997; Zhao et al., 1998). The amines alone are effective extractants, with efficacy increasing in the order of: $1j > 2j > 3j$ alkyl amines. Aromatic diluents or kerosene performed better than *n*-octanol and chloroform, apparently due to inductive electron acceptor effects of the latter solvents on amines (Chen et al., 1996; Zhao et al., 1998). A solution of NaOH (>10 pH) efficiently stripped gold from the loaded organic in a 10-min contact time (Zhao et al., 1998). Phosphorus-based organic compounds performed better in the presence of the primary amine than alone, suggesting synergistic electron-donating effects (Zhao and Wu, 1997; Zhao et al., 1998). The performance of the amine extractants was significantly improved in the presence of a trialkyl amine oxide (TRAO), which was also accounted for by the electron-donating synergism solvent. The gold complex is partitioned into the organic phase, whereas the other metals ideally remain in the aqueous phase. The organic phase may then be separated for stripping of gold with suitable strippent solution, and then after the regenerated exractant can be sent back to the extraction circuit.

4.5.4 Recovery of Gold Using Ion-Exchange Resins

The resin ion-exchange process, mostly employing strong base resins, has been widely proposed for the recovery of gold from thiosulfate leach liquors. The adsorption process can be presented as:

$$3(-^+NR_3)_2\,SO_4^{2-} + 2Au(S_2O_3)_3^{2-} = 2(-^+NR_3)_3Au(S_2O_3)_3^{2-} + 3SO_4^{2-} \qquad (4.24)$$

Despite being suitable for adsorbing the gold-thiosulfate complex, the recovery is complicated due to the presence of various sulphur species (polythionate, tetrathionate, trithionate, etc.) generated by the oxidation reactions of thiosulfate. These species get adsorbed onto the strong base resins and reduce the loading of gold by competing with gold complex ions (Fagan, 2000; O'Malley, 2001). A typical concentration of 350 mg/L trithionate and 420 mg/L tetrathionate can suppress gold adsorption from 26 to 2 kg/t of the strong base resin from a solution of 0.3 mg/L Au. An addition of 0.5 g/L Na_2SO_3 to the leach liquor in an inert atmosphere can control this detrimental effect by converting the tetrathionate back to thiosulfate. Trithionate can also be eliminated by the addition of sulphide, but the undesirable precipitation of gold sulphide is problematic.

Introducing resin-in-pulp (RIP) can also minimize the degradation of thiosulfate leach liquor; however, a relatively dilute leach liquor is required (containing 0.03–0.05 mol/L thiosulfate, 0.5–1.6 mmol/L Cu^{2+}, and 7–100 mmol/L NH_3 at pH 7–9). RIP can co-adsorb the complexes of both gold and copper, which subsequently go to selective elution. Copper can be eluted first in ammoniacal thiosulfate (100–200 g/L). Alternatively, the use of an oxygenated buffer solution of NH_4OH-$(NH_4)_2SO_4$ can preferably facilitate the recycling of eluent solution in leaching. Gold from the Cu-depleted leach solution can be eluted in the next step in a thiocyanate solution. Notably, the high cost and toxicity of thiocyanate counter-ions on the resins are undesirable (Marchbank et al., 1996) and can be replaced by a cheaper and less toxic reagent, trithionate or tetrathionate (dosage 40–200 g/L). The polythionate eluents cause less deviation in pH levels, thus minimizing osmotic shock and consequent resin attrition. The effects of trithionate and tetrathionate on gold elution are shown in Table 4.1. Further, the resin can be regenerated by flushing with a (~2 g/L) solution of sodium hydrogen sulphide, recycling both tetrathionate and trithionate into thiosulfate ions:

$$S_3O_6^{2-} + S^{2-} = 2S_2O_3^{2-} \tag{4.25}$$

$$4S_4O_6^{2-} + 2S^{2-} + 6OH^- = 9S_2O_3^{2-} + 3H_2O \tag{4.26}$$

A series of ion exchange polymers adsorbing 9.5–17.9 mg/L gold from an ammonia-thiosulfate leach liquor revealed a higher adsorption at lower pH values (Kononova et al., 2001). AV-17-10P, a trimethyl ammonium (strong base) functionalized resin could achieve a 94% adsorption efficiency at pH 6 and 85% at pH 11. However, a fair recovery of gold was also observed at pH 6 by the resins with both strong- and weak-base groups (polyfunctional resins), resins with weak-base groups, and those with amphoteric phosphonic acid – pyridine copolymers (Table 4.2). Subsequently, the elution of gold could yield >93% efficiency after 1 h operation at ambient temperature by thiourea in aqueous sulfuric acid (0.5 mol/L each). The larger capacity of strong-base resins should also make them more tolerant to low levels of competing

TABLE 4.1

Polymer Adsorbents for Au from Ammoniacal Thiosulfate Liquors

Resin Data	Type	Meq/g	Matrix	[NH$_3$]	[S$_2$O$_3$$^{2-}$]	Sorption (h)	pH	Temp (C°)	[Cu]^a	[Au]^a	Au (%)	Eluents (s)	S/R (Au, Cu)	Ref.
IRA-743	Weak base (WS)	1.0	–	7–100 mM	0.03 M	12	7–9	–	–	–	–	Cu: SO$_4^{2-}$ S$_2$O$_3^{2-}$, NH$_3$	–	Marchbank et al. (1996)
Amberlite A7	Weak base (WS)	–	Macro-porus	–	–	–	–	–	–	–	–	Au:SCN$^-$	–	
Type I/II resins	Strong base (SB)	–	–	–	–	–	–	–	–	–	–			
A500C (Purolite)	Quaternary ammonium (R$_4$N$^+$)	1.7	Macro-porus	0.1 M	0.05 M	6 × 1	8.0	60	22	1.8	99.45	Au:SCN$^-$	0.034	
A500C (Purolite)	Quaternary ammonium (R$_4$N$^+$)	1.7	Macro-porus	–	–	4	6–8	20	–	–	–	Au: S$_2$O$_6^{2-}$ + S$_2$O$_6^{2-}$	0.038	Fleming et al. (2001)
AV-17-10P, etc	Quaternary ammonium (R$_4$N$^+$)	–		0.5 M	0.5 M	5	5–11	–	–	9.5–17.9	94.2	Au: thiourea + H$_2$SO$_4$	–	Kononova et al. (2001)
Amberjet 4200 (Rohm and Hass)	R$_4$N$^+$	3.7	Gel	0.8 M	0.05 M^b	6 × 2	8–9.5	22	8.7	8.9	>99.44	Au: SCN$^-$/NO$_3^-$	2.39	O'Malley (2001)
Aurix (Henkel)	Guanidyl (Gaun)	–	–	0.05–0.1 M	25–200 mM	–	8.0–10	–	–	–	–	Au: NaOH, CN$^-$ and Sodium benzoate	–	Viring and Sierakoski (2001)
Amberjet 4200	Quaternary ammonium (R$_4$N$^+$)	3.7	Gel	0.2 M	0.05 M	–	9.5	–	10	10	99	Cu: SO$_4^{2-}$, NH$_3$, O$_2$	–	Nicol and O'Malley (2002)

(Continued)

TABLE 4.1 (*Continued*)

Polymer Adsorbents for Au from Ammoniacal Thiosulfate Liquors

Resin Data	Type	Meq/g	Matrix	[NH$_3$]	[S$_2$O$_3^{2-}$]	Sorption (h)	pH	Temp (C°)	[Cu]^a	[Au]^a	Au (%)	Eluents (s)	S/R (Au, Cu)	Ref.
Vitrokele911 (R and H)	Quaternarny ammonium (R$_4$N$^+$)	1.7	Macro-reticular									Au: NH$_4$NO$_3$(2M)		Mohansingh (2000)
IRA-400 (R and H)	Quaternarny ammonium (R$_4$N$^+$)	3.8	Gel	0.1M	1.0M	8	9–12	9.27	31.8	9.27	94.7	Au: NaCl(5M)	–	
Dowex (Dow)	Quaternarny ammonium (R$_4$N$^+$)	–	–					9.12		9.12	75.98			
Elchrom	Strong acid (SA)	–	–					9–31		9.31	9.23			

Note: Where a = ppm, b = 1.0 mM.

TABLE 4.2

Recovery of Gold by Polymer Sorbents

Adsorbent	Functional Group Affixed to Polymer	Total Exchange Capacity meq/g	Strong Base Capacity (SBC) meq/g	Percent Gold Recovery at pH = 5.8–6.1	Percent Gold Recovery at pH = 10.8–11.0
AP-100	1° & 2°-amines + ®–N + R$_3$	3.9	0.7	91.3	72.9
AP-24-10P	®–N(CH$_2$CH$_2$CH$_2$CH$_3$)$_2$ & ® –(CH$_3$)$_2$N + Ph	4.1	1.3	86.5	47.9
AP-2–12P	®–(CH$_3$)$_2$N + CH$_2$N(CH$_3$)$_2$	3.7	1.1	90.4	65.8
AV-17-10P	®–N$^+$(CH$_3$)$_3$	4.4	4.1	94.2	85.4
AN-106-7P	®–NH(CH$_2$)$_2$NH(CH$_2$)$_2$NH$_2$	8.9	–	88.1	39.0
AN-85-10P	®–NHCH$_2$CH$_2$NH$_2$	6.2	–	86.5	51.0
ANFK-5 (No pores)	®–PO(ONa)$_2$+{N(CH$_3$)$_2$}$_n$ +Poly(vinylpyridine)	Unknown	–	87.5	54.2

Note: Conditions: 5 h contact time, 0.5 M Na$_2$S$_2$O$_3^{2-}$, 0.5 M NH$_3$, ambient temperature, gold 9.5–17.9 ppm, ®, resin backbone.

anions (Nicol and O'Malley, 2002). However, Minix, a strong-base resin with excellent selectivity for aurocyanide, performed as poorly as the weak-base/polyamine resins (at pH 8). It has been noted that copper concentration in the pulp increases as the gold displaces it from the resin; hence due to a gradual formation of the competing ions trithionate and tetrathionate in the liquor, the resin contact time must be minimized (Nicol and O'Malley, 2002; O'Malley, 2001). In addition to gold, the thiosulfate complexes of other metals such as lead, copper, zinc, and silver are adsorbed onto strong-base resins from the leach liquors (Fagan, 2000; O'Malley, 2001). This reduces the loading capacity for gold adsorption, and unless these can be separately eluted, these metals will contaminate the final gold eluate. The proposed affinity order for adsorption of thiosulfate complexes onto strong-base resins, based on mixed metal adsorption tests, was reported to be: Au > Pb > Ag > Cu > Zn. Although kinetically slower to adsorb, the presence of these anions in leach liquors restricts the maximal gold recovery that can be achieved. This may become a major problem for the more aggressive leaching operations, where considerable thiosulfate is consumed during leaching and hence polythionates are abundant in the liquor. In a claim by Henkel (Australia) the application of guanidine to functionalized polystyrene resin (*Aurix*) for the recovery of gold from liquors containing thiosulfate (25–200 mM) and ammonia (>50 mM) at pH 8–10 (Virnig and Sierakoski, 2001) permits the protonated form of resin, due to a high pK_a value of 13.5. The proponents claim that gold can be stripped from the sorbent using aqueous NaOH (pH > 11), with the optional additives of NaCN and/or a carboxylic acid such as sodium benzoate to facilitate more efficient elution.

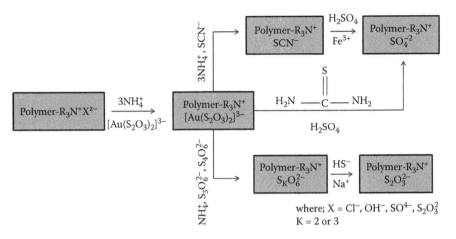

FIGURE 4.8
Gold adsorption by a strong-base resin followed by various elution options.

Gold adsorption by a strong-base resin followed by various elution options is schematically shown in Figure 4.8.

4.6 Recyclability of Gold-Depleted Thiosulfate Solutions

The recyclability of thiosulfate solution strongly depends on the metal recovery technique employed, which may cause significant degradation of the liquor by oxidation, reduction, or contamination (Awadalla and Ritcey, 1991; Benedetti and Boulègue, 1991), whereas ammonia can be stripped from metal-depleted liquor (tailings) by exploiting the significant volatility (Marchbank et al., 1996). Byproducts and tailings from thiosulfate processing should consist primarily of low toxicity metal hydroxides, oxides, sulphates, polythionates, polysulfides, and/or insoluble sulphides, although pilot studies to date have not directly addressed waste management. There are several reversible reactions in which thiosulfate is either consumed or regenerated, some of which are vital in leaching by recycling various breakdown products and the regeneration of thiosulfate ions, as shown in Equations 4.27–4.44 where each reaction formulates the thiosulfate as product (Roy and Trudinger, 1970; Byerley et al., 1975; Perez and Galaviz, 1987; Zipperian et al., 1988; Hu and Gong, 1991; Xu and Schoonen, 1995; Bean, 1997; Fleming, 1998):

$$SO_3^{2-} + 2OH^- + S_4O_6^{2-} = 2S_2O_3^{2-} + SO_4^{2-} + H_2O \qquad (4.27)$$

$$3SO_3^{2-} + 2S^{2-} + 3H_2O = 2S_2O_3^{2-} + 6OH^- + S^0 \tag{4.28}$$

$$SO_3^{2-} + S_5O_6^{2-} = S_2O_3^{2-} + S_4O_6^{2-} \tag{4.29}$$

$$SO_3^{2-} + S_4O_6^{2-} = S_2O_3^{2-} + S_3O_6^{2-} \tag{4.30}$$

$$2S_5O_6^{2-} + 6OH^- = 5S_2O_3^{2-} + 3H_2O \tag{4.31}$$

$$2S_5O_6^{2-} + 6OH^- = 5S_2O_3^{2-} + 3H_2O \tag{4.32}$$

$$4S_4O_6^{2-} + 6OH^- = 5S_2O_3^{2-} + 2S_3O_6^{2-} + 3H_2O \tag{4.33}$$

$$2S_3O_6^{2-} + 6OH^- = S_2O_3^{2-} + 4SO_3^{2-} + 3H_2O \tag{4.34}$$

$$4S_4O_6^{2-} + H_2S = 2S_2O_3^{2-} + S^0 + 2H^+ \tag{4.35}$$

$$2S_2O_4^{2-} = S_2O_3^{2-} + S_2O_5^{2-} \tag{4.36}$$

$$2S^{2-} + 4SO_4^{2-} + 8H^+ + 8e^- = 3S_2O_3^{2-} + 6OH^- + H_2O \tag{4.37}$$

$$2S^{2-} + 2SO_2 + 2HSO_3^- = 3S_2O_3^{2-} + H_2O \tag{4.38}$$

$$2S^{2-} + 3SO_2 + SO_3^{2-} = 3S_2O_3^{2-} \tag{4.39}$$

$$2HS^- + 4HSO_3^- = 3S_2O_3^{2-} + 3H_2O \tag{4.40}$$

$$S_6O_6^{2-} + 3SO_3^{2-} = 3S_2O_3^{2-} + S_3O_6^{2-} \tag{4.41}$$

$$S^0 + SO_3^{2-} = S_2O_3^{2-} \tag{4.42}$$

$$S_{(x)}^{2-} + SO_3^{2-} = S_2O_3^{2-} + S_{(x-1)}^{2-} \tag{4.43}$$

$$S_3O_6^{2-} + S^{2-} = 2S_2O_3^{2-} \tag{4.44}$$

Aiming the regeneration of decomposed thiosulfate and lixiviating the refractory MnO_2, sulphite addition has been suggested by different researchers (Johnson and Bhappu, 1969; Flett et al., 1983; Hemmati et al., 1989; Lulham and Lindsay, 1991; Langhans et al., 1992; Groudev et al., 1996; Guerra and Dreisinger, 1999). But the actual benefits of sulphite addition are questionable, due to ready oxidation of sulphite by Cu^{2+}, which produces Cu^+, sulphate, and dithionate ions (Aylmore, 2001). Augmentation with excess sulphate to enhance thiosulfate stability (Hemmati et al., 1989; Hu and Gong, 1991; Gong et al., 1993) involving an eight electron redox reaction for reduction of sulphate to thiosulfate may not be feasible (shown in Equation 4.37). Apart from sulphide and sulphate, the breakdown products of thiosulfate are not known to form stable complexes with the metal ions of interest (Smith and Martell, 1976; Aylmore, 2001). Metal sulphide complexes are generally sparingly soluble, while sulphate has negligible chelating ability, and complexes incorporating other polythionate $(S_xO_y^{2-})$ ligands are overwhelmed by the abundant thiosulfate ions. A number of authors have reported the *in situ* synthesis of thiosulfate ions from sulphoxy compounds or ions during a controlled oxidative leaching (Genik-Sas-Berezowsky et al., 1978; Groves and Blackman, 1995; Chen et al., 1996). As a byproduct of the destruction of a sulphide matrix, the oxidation of native sulphur or sulphides may be the cheapest source of lixiviant generation (Equations 4.42, 4.43 and 4.45–4.48) (Chen et al., 1996; Bean, 1997).

$$4S^0 + 6OH^- = S_2O_3^{2-} + 2S^{2-} + 3H_2O \qquad (4.45)$$

$$2NH_3 + SO_2 + S^0 + H_2O = 2(NH_4)_2S_2O_3 \qquad (4.46)$$

$$2(NH_4)_2S + 2SO_2 + O_2 = 2(NH_4)_2S_2O_3 \qquad (4.47)$$

$$S_8 + 8NaOH = 2(Na_2S_2O_3) + 2NaS_xH + H_2O \qquad (4.48)$$

The reaction mechanism that permits such transformation appears to involve the attack on elemental sulphur by transitory polysulfide species (i.e., NaS_xH) (Bean, 1997). The sulphur dioxide produced in an ore roasting step may also be used to generate thiosulfate lixiviant (Aylmore and Muir, 2001). Recovering harmful sulphurous matter in this fashion also has the advantage of minimizing the environmental impact of the operation. However, recycling and/or *in situ* generation of thiosulfate has yet to be implemented at a significant scale.

4.7 Environmental Impact, Limitations, and Challenges for Thiosulfate Leaching of Gold

Although thiosulfate is regarded as nontoxic, it decomposes to polythionates and sulphate with oxygen, or to toxic sulphide ions at reducing conditions; hence, the uncontrolled discharge poses environmental problems of deoxygenation of water streams. The formation of decomposition products can be avoided by oxidation of thiosulfate to sulphate prior to discharge, but the oxidation cost is higher than the cyanide oxidation (Lee and Srivastava, 2016). Recycling of thiosulfate solution is a possibility; however, building up of polythionates is another issue, making the use of a minimal reagent a necessity. Moreover, ammonia also poses environmental issues both in gaseous and liquid form. The threshold limiting value for gaseous ammonia in air is 14 mg/m^3 (Gos and Rubo, 2000), whereas its toxicity in water is similar to chlorine; hence, strict precautions are needed for thiosulfate leaching.

Thiosulfate leach reactions are less favourable than gold cyanidation (Lee and Srivastava, 2016) and therefore are consumed in high amounts to achieve the equivalent rates for gold leaching. A thiosulfate leach solution requires a concentration of 5–20 vs. 0.25–1 g/L cyanide in solution. A higher consumption of thiosulfate partially offsets its significantly lower cost (approximately one-fifth the cost of cyanidation). A poor affinity to adsorb gold-thiosulfate complex onto carbon is negating the use of conventional processes of carbon-in-pulp or carbon-in-leach. Instead, the applicability of RIP technology in gold recovery from leach pulp by strong-base resin should be more prominent. RIP requires mild leaching conditions, as the strong thiosulfate concentration may raise the competitive adsorption on resin sites by decomposition products, such as polythionate.

References

Abbruzzese, C., Fornari, P., Massidda, R., Veglio, F., Ubaldini, S. 1995. Thiosulfate leaching for gold hydrometallurgy. *Hydrometallurgy*. 39(1): 265–276.

Anthony, E.Y., Williams, P.A. 1994. Thiosulfate complexing of platinum group elements: Implications for supergene geochemistry. In: Alpers, C.L., Blowes, D.W. (Eds), *Environmental Geochemistry of Sulfide Oxidation*. ACS Symposium Series, American Chemical Society, Washington, DC.

Awadalla, F., Ritcey, G. 1991. Recovery of gold from thiourea, thiocyanate, or thiosulfate solutions by reduction-precipitation with a stabilized form of sodium borohydride. *Separation Science and Technology*. 26(9): 1207–1228.

Aylmore, M.G. 2001. Treatment of a refractory gold—Copper sulphide concentrate by copper ammoniacal thiosulfate leaching. *Minerals Engineering*. 14(6): 615–637.

Aylmore, M.G., Muir, D.M. 2001. Thiosulfate leaching of gold—A review. *Minerals Engineering*. 14(2): 135–174.

Bean, S.L. 1997. Thiosulfates. In: Kroschwitz, J.I. (Ed.), *Kirk-Othmer Encyclopedia of Chemical Technology*. Wiley, New York, pp. 51–68.

Benedetti, M., Boulègue, J. 1991. Mechanism of gold transfer and deposition in a supergene environment. *Geochimica et Cosmochimica Acta*. 55(6): 1539–1547.

Berezowsky, R., Sefton, V. 1979. Recovery of gold and silver from oxidation leach residues by ammoniacal thiosulfate leaching. In: *AIME Annual Meeting*, New Orleans, LA, pp: 102–105.

Brent Hiskey, J., Atluri, V.P. 1988. Dissolution chemistry of gold and silver in different lixiviants. *Mineral Processing and Extractive Metallurgy Review*. 4(1–2): 95–134.

Briones, R., Lapidus, G. 1998. The leaching of silver sulphide with the thiosulfate–ammonia–cupric ion system. *Hydrometallurgy*. 50(3): 243–260.

Byerley, J.J., Fouda, S.A., Rempel, G.L. 1973. Kinetics and mechanism of the oxidation of thiosulfate ions by copper (II) ions in aqueous ammonia solution. *Journal of the Chemical Society, Dalton Transactions*. 889–893. DOI:10.1039/DT9730000889.

Byerley, J.J., Fouda, S.A., Rempel, G.L. 1975. Activation of copper (II) ammine complexes by molecular oxygen for the oxidation of thiosulfate ions. *Journal of the Chemical Society, Dalton Transactions*. 1329–1338. DOI:10.1039/DT9750001329

Chen, J., Deng, T., Zhu G., Zhao, J. 1996. Leaching and recovery of cold in thiosulfate based system-a research summary at ICM. *Transactions of the Indian Institute of Metals*. (6): 841–849.

Deschenes, G., Ritcey, G.M. 1990. Recovery of gold from aqueous solutions. Google Patents.

Fagan, P. 2000. Personal communication: Report on the Ballarat gold forum—Gold processing in the 21st century: An international forum, Ballarat, VIC.

Fleming C.A. 1998. The potential role of anion exchange resins in the gold industry. EPD Congress 1998. The Minerals, Metals and Materials Society, Warrendale, PA, pp. 95–117.

Fleming, C.A., McMullen, J., Thomas, K.G., Wells, J.A. 2001. Recent advances in the development of an alternative to the cyanidation process: Thiosulfate leaching and resin in pulp. *Minerals and Metallurgical Processing*. 20: 1–9.

Flett, D., Derry, R., Wilson, J. 1983. Chemical study of thiosulfate leaching of silver sulphide. *Transactions of the Institution of Mining and Metallurgy Section C-Mineral Processing and Extractive Metallurgy*. 92(DEC): C216–C223.

Gallagher, N., Hendrix, J., Milosavljević, E., Nelson, J. 1989. The affinity of carbon for gold complexes: Dissolution of finely disseminated gold using a flow electrochemical cell. *Journal of the Electrochemical Society*. 136(9): 2546–2551.

Genik-Sas-Berezowsky, R.M., Sefton, V.B., Gormely, L.S. 1978. Recovery of precious metals from metal sulphides. U.S. Patent No. 4, 070, 182. U.S. Patent and Trademark Office, Washington, DC.

Gmelin, L.E. 1965. Copper thiosulfate complexes. In: *Gmelins Handbuch der Anorganischen Chemie*. Verlag Chemie, Berlin, Germany, pp. (B1) 591, (B593) 998–1003, 1414–1415.

Gmelin, L.E. 1969. Mercury thiosulfate complexes. In: *Gmelins Handbuch der Anorganischen Chemie*. Verlag Chemie, Berlin, Germany, pp. (B1) 52–55, (B53) 1030–1031, (B1034) 1406–1409.

Gmelin, L.E. 1972. Lead thiosulfate complexes. In: *Gmelins Handbuch der Anorganischen Chemie*. Verlag Chemie, Weinheim, Germany, pp. (B1) 364–365, (C) 594–600.

Gmelin, L.E. 1973. Silver thiosulfate complexes. In: *Gmelins Handbuch der Anorganischen Chemie*. Verlag Chemie, Weinheim/Bergstrasse, Germany, pp. (B3) 110–133.

Gong, Q., Hu, J., Cao, C. 1993. Kinetics of gold leaching from sulphide gold concentrates with thiosulfate solution. *Transactions of the Nonferrous Metals Society of China*. 3(4): 30–36.

Gos, S., Rubo, A. 2000. Alternative lixiviants for gold leaching. A comparison. In: *Randol Gold & Silver Forum*, Randol International Ltd., Golden, CO, pp. 271–281.

Groudev, S.N., Spasova, I.I., Ivanov, I.M. 1996. Two-stage microbial leaching of a refractory gold-bearing pyrite ore. *Minerals Engineering*. 9(7): 707–713.

Groves, W.D., Blackman, L. 1995. Recovery of precious metals from evaporate sediments. U.S. Patent No. 5, 405, 430. U.S. Patent and Trademark Office, Washington, DC.

Guerra, E., Dreisinger, D.B. 1999. A study of the factors affecting copper cementation of gold from ammoniacal thiosulfate solution. *Hydrometallurgy*. 51(2): 155–172.

Hemmati, M., Hendrix, J., Nelson, J., Milosavljevic, E. 1989. In: *Extraction Metallurgy '89 Symposium*. Institute of Mining and Metallurgy, London, pp. 665–678.

Hu, J., Gong, Q. 1991. Substitute sulphate for sulphite during extraction of gold by thiosulfate solution. In: *Randol Gold Forum '91*, Cairns, QLD, pp. 333–336.

Hubin, A., Vereecken, J. 1994. Electrochemical reduction of silver thiosulfate complexes part I: Thermodynamic aspects of solution composition. *Journal of Applied Electrochemistry*. 24(3): 239–244.

Jiang, T., Chen, J., Xu, S. 1993. A kinetic study of gold leaching with thiosulfate. In: Hiskey, J.B., Warren, G.W. (Eds), *Hydrometallurgy, Fundamentals, Technology and Innovations*. AIME, Society for Mining, Metallurgy, & Exploration, Littleton, CO, pp. 119–126.

Johnson, P.H., Bhappu, R.B. 1969. Chemical mining: A study of leaching agents, Circular 99. State Bureau of Mines and Mineral Resources, New Mexico Institute of Mining and Technology, Socorro, NM.

Johnson, J.A., Davis, J. O. 1973. Effects of a specific competitive antagonist of angiotensin II on arterial pressure and adrenal steroid secretion in dogs. *Circulation Research*. 32(5): I–159.

Kononova, O., Kholmogorov, A., Kononov, Y., Pashkov, G., Kachin, S., Zotova, S. 2001. Sorption recovery of gold from thiosulfate solutions after leaching of products of chemical preparation of hard concentrates. *Hydrometallurgy*. 59(1): 115–123.

Langhans, J., Lei, K., Carnahan, T. 1992. Copper-catalysed thiosulfate leaching of low-grade gold ores. *Hydrometallurgy*. 29(1–3): 191–203.

Lee, J.-C., Srivastava, R.R. 2016. Leaching of gold from the spent/end-of-life mobile phone-PCBs. In: Sabir, S. (Ed.), *The Recovery of Gold from Secondary Resources*. Imperial College Press, London, UK, pp: 7–56.

Li, J., Miller, J., Le-Vier M., Wan, R. 1995. *The Ammoniacal Thiosulfate System for Precious Metal Recovery*. Society for Mining, Metallurgy, and Exploration, Inc., Littleton, CO.

Livingstone, S.E. 1965. Metal complexes of ligands containing sulphur, selenium, or tellurium as donor atoms. *Quarterly Reviews, Chemical Society*. 19(4): 386–425.

Lulham, J., Lindsay, D. 1991. Separation process. International Patent WO 91/11539. Davy McKee (Stockton) Limited, Cleveland, Great Britain.

Marchbank, A.R., Thomas, K.G., Dreisinger, D., Fleming, C. 1996. Gold recovery from refractory carbonaceous ores by pressure oxidation and thiosulfate leaching. Google Patents.

Meggiolaro, V., Niccolini, G., Miniussi, G., Stefanelli, N., Trivellin, E., Llinas, R., Ramirez, I., BaccelleScudeler, L., Omenetto, P., Primon, S., Visona, D., Abbruzzese, C., Fornari, P., Massidda, R., Piga, L., Ubaldini, S., Ball, S., Monhemius, A.J. 2000. Multidisciplinary approach to metallogenic models and types of primary gold concentration in the Cretaceous arc terranes of the Dominican Republic. *Transactions of the Institutions of Mining and Metallurgy: Section B*. 109: b95–b104 (May–Aug).

Meng, X., Han, K.N. 1993. The dissolution b›aviour of gold in ammoniacal solutions. In: Hiskey, JB, Warren, GW (Eds), *Hydrometallurgy Fundamentals, Technology and Innovations*. Society for Mining, Metallurgy and Exploration, Englewood, CO, pp. 205–221.

Mohansingh, R. 2000. Adsorption of gold from gold copper ammonium thiosulfate complex onto activated carbon and ion exchange resins. Master's Thesis, University of Nevada, Reno, NV, May 2000.

Mountain, B.W., Wood, S.A. 1988. Solubility and transport of platinum-group elements in hydrothermal solutions: Thermodynamic and physical chemical constraints. In: Prichard, H.M., Potts, P.J., Bowles, J.F.W., Cribb, S.J. (Eds), *Geo-Platinum 87*. Springer, Dordrecht, pp: 57–82.

Muir, D.M., La Brooy, S.R., Deng, T., Singh, P. 1993. The mechanism of the ammonia-cyanide system for leaching copper-gold ores. In: Hiskey, J.B., Warren, G.W. (Eds), *Hydrometallurgy: Fundamentals, Technology and Innovation*. AIME, Society for Mining, Metallurgy, & Exploration, Littleton, CO, pp. 191–204.

Naito, K., Shiʾ, M.-C., Okabe, T. 1970. The chemical bʾaviour of low valence sulphur compounds. V. Decomposition and oxidation of tetrathionate in aqueous ammonia solution. *Bulletin of the Chemical Society of Japan*. 43(5): 1372–1376.

Nicol, M.J., O'Malley, J. 2002. Recovering gold from thiosulfate leach pulps with ion-exchange. *Journal of the Minerals, Metals, and Materials Society*. 54(10): 44–46..

Novakovskii, M.S., Ryazantseva, A.P. 1955. Cadmium complexes with thiosulfate. *Trudy Khim. Fak.* 54 (12): 277–281 (In Russian).

O'Malley, G. 2001. The elution of gold from anion exchange resins. Australian Patent. WO 123626.

Pedraza, A., Villegas, I., Freund, P., Chornik, B. 1988. Electro-oxidation of thiosulfate ion on gold: Study by means of cyclic voltammetry and auger electron spectroscopy. *Journal of Electroanalytical Chemistry and Interfacial Electrochemistry*. 250(2): 443–449.

Perez, A.E., Galaviz, H.D. 1987. Method for recovery of precious metals from difficult ores with copper-ammonium thiosulfate. U.S. Patent No. 4, 654, 078. U.S. Patent and Trademark Office, Washington, DC.

Plimer, I.R., Williams, P.A. 1988. New mechanisms for the mobilization of the platinum-group elements in the supergene zone. In: *Geo-Platinum 87*. Elsevier Applied Science Publishers, London, pp. 83–92.

Roy, A.B., Trudinger, P.A. 1970. The Biochemistry of Inorganic Compounds of Sulphur. Cambridge University Press, Cambridge, UK.

Ryabchikov, D.I. 1943. On the structure of dithiosulphatoplatinite. In: *Comptes Rendus (Doklady) de l'Academie des Sciences de l'URSS*. 41(5): 208–209.

Shiʾ, M.-C., Otsubo, H., Okabe, T. 1965. The chemical bʾaviour of low valence sulphur compounds. I. Oxidation of elemental sulphur with compressed oxygen in aqueous ammonia solution. *Bulletin of the Chemical Society of Japan*. 38(10): 1596–1600.

Smith, R.M., Martell, A.E. 1976. Inorganic ligands. In: Martell, A. (Ed.), *Critical Stability Constants*. Springer, Boston, pp. 1–129.

Ter-Arakelyan, K. 1984. On technological expediency of sodium thiosulfate usage for gold extraction from raw material. *Izvestiya Vysshikh Uchebnykh Zavedenii, Tsvetnaya Metallurgia*. 5: 72–76.

Tozawa, K., Inui, Y., Umetsu, Y. 1981. Dissolution of gold in ammoniacal thiosulfate solution. In: *AIME 110th Annual Meeting*. A81-25, pp. 1–12.

Tykodi, R. 1990. In praise of thiosulfate. *Journal of Chemical Education*. 67(2): 146.

Tyurin, N.G., Kakowski, I.A. 1960. Bʾaviour of gold and silver in oxidising zine of sulphide deposit. Izu., *Buz. Tsyv. Metallurgy* 2: 6–13 (Russian).

　　　　　　　　　　　　　　　　　　　　Gold Metallurgy and the Environment

Umetsu, Y., Tosawa, K. 1972. Dissolution of gold in ammoniacal sodium thiosulfate solution. *Bulletin of the Research Institute of Mineral Dressing and Metallurgy.* 28(1): 97–104.

Uri, N. 1947. Thiosulfate complexes of the tervalent metals, iron, aluminium, and chromium. *Journal of the Chemical Society* (Resumed). 335–336.

Vasil'ev, A., Toropova, V.F., Busygina, A.A. 1953. The use of ion exchange for the separation of copper, cadmium and zinc from thiosulfate solutions. *Uch. Zap. Kazausk. Un-ta.* 113 (118): 91–102.

Virnig, M.J., Sierakoski, J.M. 2001. Ammonium thiosulfate complex of gold or silver and an amine. Google Patents.

Vlassopoulos, D., Wood, S.A. 1990. Gold speciation in natural waters: I. Solubility and hydrolysis reactions of gold in aqueous solution. *Geochimica et Cosmochimica Acta.* 54(1): 3–12.

Von Michaelis, H. 1987. Thiosulfate leaching of gold and silver. Randol phase IV report, Randol International, Golden, CO, pp. 4876–4885.

White, D. 1900. *The Stratigraphic Succession of the Fossil Floras of the Pottsville Formation in the Southern Anthracite Coal Field, Pennsylvania.* US Government Printing Office, Washington, DC.

Wilkinson, G., Gillard, R.D. 1987. Middle transition elements. In: Wilkinson, G., Gillard, R.D., McCleverty, J.A. (Eds), *Comprensive Coordination Chemistry: The Synthesis, Reactions, Properties and Applications of Coordination Compounds.* Pergamon Press, Oxford.

Williamson, M.A., Rimstidt, J.D. 1993. The rate of decomposition of the ferric-thiosulfate complex in acidic aqueous solutions. *Geochimica et Cosmochimica Acta.* 57(15): 3555–3561.

Xu, Y., Schoonen, M.A. 1995. The stability of thiosulfate in the presence of pyrite in low-temperature aqueous solutions. *Geochimica et Cosmochimica Acta.* 59(22): 4605–4622.

Zhang, H., Dreisinger, D.B. 2002. The adsorption of gold and copper onto ion-exchange resins from ammoniacal thiosulfate solutions. *Hydrometallurgy.* 66(1): 67–76.

Zhao, J., Wu, Z. 1997. Extraction of gold from thiosulfate solutions with alkyl phosphorus esters. *Hydrometallurgy.* 46: 363–372.

Zhao, D., Feng, J., Huo, Q., Melosh, N., Fredrickson, G.H., Chmelka, B.F., Stucky, G.D. 1998. Triblock copolymer syntheses of mesoporous silica with periodic 50 to 300 angstrom pores. *Science.* 279(5350): 548–552.

Zipperian, D., Raghavan, S., Wilson, J.P. 1988. Gold and silver extraction by ammoniacal thiosulfate leaching from a rhyolite ore. *Hydrometallurgy.* 19(3): 361–375.

5

Thiourea Leaching of Gold

Sadia Ilyas*, Muhammad Ahmed Mohsin*, and Jae-chun Lee†

5.1 Development of Gold Leaching in Thiourea Solution

The use of thiourea as an alternative lixiviant to gold cyanidation for processing gold-bearing ores or concentrates was first developed by Soviet metallurgists in the 1940s (Plaskin and Kozhukhova, 1941). Since then, considerable research has been directed toward the use of thiourea; however, the interest for thiourea leaching of gold mainly came during the 1980s and 1990s to establish a parallel process (Groenewald, 1977). Canmet, Mint–, Newmont Mining, and Barrick Gold looked closely at thiourea as a potent lixiviant for processing the refractory ores (van Staden and Laxen, 1989; Tremblay et al., 1996). But it recently lost its attraction; the carcinogenic properties of thiourea cannot stand as an alternative to cyanidation. However, laboratory research is being carried out from time-to-time (Örgül and Atalay, 2000, 2002; Gönen, 2003; Zhang et al., 2012). Low sensitivity to base metals (Cu, Pb, Zn, As, etc. mainly as impurities) and residual sulphur and high recovery from pyrites and carbonaceous ores are some advantages of thiourea processing, which were considered as an alternative to cyanidation.

5.2 Influential Factors for Extracting Gold in Thiourea Solution

Thiourea is commonly unstable in alkaline media, whose well-known tautomeric forms is shown in Figure 5.1. The chemistry of thiourea is simpler than thiosulfate complexation with gold, forming a cationic complex of gold.

* Mineral and Material Chemistry Lab, Department of Chemistry, University of Agriculture Faisalabad, Pakistan.
† Minerals Resources Research Division, Korea Institute of Geoscience and Mineral Resources, Daejeon, South Korea.

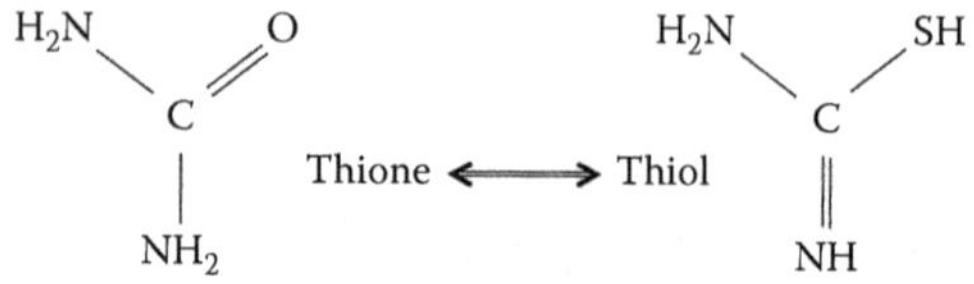

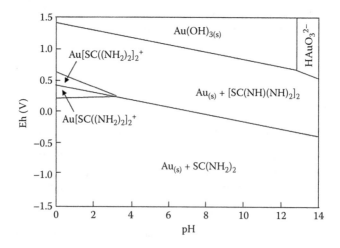

FIGURE 5.1
The two well-known tautomeric forms of thiourea.

The formation of gold complexes in thiourea solution can be better represented by the Eh-pH diagram, shown in Figure 5.2. As can be seen from Figure 5.2, thiourea is unstable and decomposes in solutions that do not have an acid pH. Hence, the leaching of gold is most prominently performed in the pH range from 1–2 to form the only common existing cationic species of gold, $Au[SC(NH_2)_2]_2^+$ (Munoz and Miller, 2000; Brent Hiskey and Atluri, 1988). Basically, the electro-chemical nature of the reaction, yielding up to 99% of gold can be presented as (Li and Miller, 2006):

$$Au + 2SC(NH_2)_2 = Au[SC(NH_2)_2]_2^+ + e^- \quad E = 0.38 \text{ V} \tag{5.1}$$

5.2.1 Effect of Oxidants/Ferric Ions

Various oxidants like oxygen, ozone, sodium peroxide, hydrogen peroxide, ferric ions, dichromate, and manganese dioxide are employed to enhance the leaching of gold in thiourea solution. Amongst all, the ferric ions are found to be the most effective, more preferably in an acidic sulphate solution than

FIGURE 5.2
Eh-pH diagram for formation of gold complexes in thiourea solution.

the chloride/nitrate media (Plaskin and Kozhukeva, 1960; Songina et al., 1971). The reaction for gold leaching in thiourea and ferric ion solutions can be written as:

$$Au + 2SC(NH_2)_2 + Fe^{3+} = Au[SC(NH_2)_2]_2^+ + Fe^{2+} \tag{5.2}$$

In the presence of redox couple Fe^{3+} and Fe^{2+}, gold leaching in thiourea solution has been found to be up to four times faster than oxidation by air purging (Huyhua et al., 1989). Deschenes and Ghali (1988) demonstrated the effect of ferric ion as an oxidizing agent during leaching of gold from chalcopyrite concentrate. Without any oxidizing agent, the thiourea solution provides an extraction of 80% gold in 8 h. An addition of 2.0 g/L of this oxidant slightly improves the initial leaching kinetics but results in no real improvement in gold extraction. Increasing concentration of the oxidant to 5.0 g/L could increase leaching kinetics and resulted in extraction of >93% gold in 7 h leaching. However, a further increase in Fe^{3+} ion (8.0 g/L) decreases the gold extraction. Studies revealed a lower dosage of ferric ions is not adequate to oxidize the metal ions; whereas a higher dosage oxidizes thiourea itself to S^{2-}, S^0 and formamidine disulphide which can suppress the leaching efficiency of gold. The oxidative degradation of thiourea into several degradation products can be written as:

$$2SC(NH_2)_2 + Fe^{3+} = [SCN_2H_3]_2 + 2Fe^{2+} + 2H^+ \tag{5.3}$$

$$[SCN_2H_3]_2 = SC(NH_2)_2 + NH_2CN + S^0 \tag{5.4}$$

$$SC(NH_2)_2 + Fe^{3+} + SO_4^{2-} = [FeSO_4 \cdot SC(NH_2)_2]^+ \tag{5.5}$$

Gold leaching in acid thiourea solutions from an oxidized gold–copper ore carried out by McInnes et al. (1989) revealed the requirement of a prior removal of copper using a dilute acid leaching for yielding an improved gold extraction. The optimum conditions for gold leaching were pH 1.5, redox potential over 200 mV, and 0.4% thiourea, yielding a 90% efficiency in 2 h leaching with consumption of 25 kg/t thiourea. In their study, Lodeishchikov et al. (1968) suggested a preliminary washing of ore concentrates with a dilute solution of H_2SO_4 before leaching the gold could achieve a higher yield. Similar observations for gold-thiourea leaching from cuprous gold-bearing ores, finely disseminated gold ore, and concentrates have been reported (Lodeishchikov et al., 1972; Pyper and Hendrix, 1981).

Tremblay et al. (1996) studied gold leaching from auriferous sulphide mineral via percolation with acidic thiourea solution and use of H_2O_2 and ferric sulphate as the oxidants. The redox potential of 0.42–0.45 V against SHE and a pre-washing ore by dilute H_2SO_4 could limit the lixiviant consumption, yielding 76% gold using H_2O_2 (consuming 1.6 kg/t) and 83% gold with ferric sulphate (consuming 2.4 kg/t).

5.2.2 Effect of pH

As can be seen from the Eh-pH diagram in Figure 5.2, the predominant area for the formation of stable soluble species of gold, $Au[SC(NH_2)_2]_2^+$ in thiourea solution is small and exists below pH 4.0. The parametric influence of solution pH (investigated using 10 g/L thiourea, 5 g/L Fe^{3+} ion, temperature 25°C, pulp density 10%, and time 7h) showed a decreasing trend for gold leaching with a change in pH from higher acidic to a lower acidic solution. The stability of thiourea at lower acidic conditions by hydrolysis decomposition (as shown in the next reaction) is another reason for limiting the leaching pH < 4.

$$SC(NH_2)_2 + H_2O = (NH_2CONH_2) + H_2S \qquad (5.6)$$

Notably, both products of the previous reaction are undesirable in gold leaching as the presence of urea may decrease the leaching kinetics due to its passivation onto surface, and H_2S may cause a reverse solubility of gold from the leach liquor (Sparrow and Woodcock, 1995). As seen in Figure 5.3, an increase in pH could remarkably decrease the leaching efficiency of gold. A remarkable decline in gold leaching from 93.2% to 80% is seen with a solution of pH 1.0–4.0, respectively (Deschenes and Ghali, 1988). The system of gold-thiourea leaching is therefore not simple and its potential application at commercial scale is contingent upon the control of various parameters (Ilyas and Lee, 2014).

5.2.3 Effect of Thiourea Concentration

To maintain a proper dosage of lixiviant, commonly a 0.13 M thiourea, 5 g/L ferric ion concentration, the potential of solution in the range of 400–450 mV

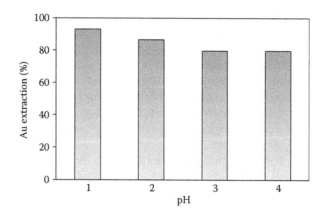

FIGURE 5.3
Effect of pH on the efficiency of gold leaching in thiourea solution.

and pH in between 1 and 2 has been found to be optimal. The effect of thiourea concentration from 5 to 30 g/L along with a 5.0 g/L Fe^{3+} ion showed only 50% gold leaching at a lower thiourea concentration (Deschenes and Ghali, 1988). The maximum 93.4% gold extraction could be achieved with a 10 g/L thiourea solution, which comes to 91.8% and 88.0% with 20 and 30 g/L thiourea, respectively in the leaching solution. Increasing thiourea concentration up to 10 g/L produces an increase in the leaching kinetics of gold, more favourable at the initial 30 min; however, some report a decrease in gold leaching >20 g/L thiourea (Groenewald, 1977). Since thiourea is less stable at higher concentrations and easily decomposes to the elemental sulphur, this causes passivation and decreases the gold extraction in the thiourea solution (Plaskin and Kozhukhova, 1960; Groenewald, 1976; Groenewald, 1977). Extraction of gold from a Turkish oxide gold ore in acidic solution of thiourea–ferric sulphate solution, investigated by Celik (2004), has suggested the addition of $Na_2S_2O_5$ (2 kg/t) inhibits the decomposition of thiourea to obtain >94% gold.

5.2.4 Effect of Temperature

Temperature plays an important role for the enhancement of leaching kinetics; hence, plenty of works have been carried out in this regard. Leaching of chalcopyrite concentrate with a 10 g/L thiourea and 5 g/L Fe^{3+} ion indicated 80.4% of efficiency at 25°C, which improves to 90.7% with increasing the temperature up to 40°C in 30 min and further to 93.5% in the next 210 min. A greater increase in temperature (≥60°C) causes a declined efficacy of leaching (~88%).

The effect can be corroborated with the thiourea decomposition at higher temperature by forming colloidal sulphur in the presence of iron (shown in Equations 5.2 and 5.3), which retards the leaching reaction through passivation onto gold surfaces (Groenewald, 1976). A temperature effect on leaching can be corroborated to the chemically-controlled reaction (Habashi, 1969), which is expected to be an electro-chemically driven reaction, as shown in Figure 5.4. The cathodic half-cell reaction for reducing the formamidine disulphide (in Figure 5.4) seems like the proton ions may participate in the rate-limiting reaction. But it has been found that pH does not change with thiourea decomposition, indicating that the formamidine disulphide exists in protonated form instead of the neutral molecule (Li and Miller, 1997, 2002). The presence of Fe^{3+} ions in the bulk thiosulfate solution catalyses the electrochemical reactions to facilitate the oxidation of metallic gold to aurous (Au^+) and silver to argentous (Ag^+) ions, as shown in Figure 5.4 (for Au^0 to Au^+).

The addition of the reducing reagent (SO_2) into acidic thiourea solution selectively reduces the formamidine disulphide to thiourea and helps keep the oxidized part of the initial thiourea to ~50% in the leaching process (Schulze, 1984). An addition of SO_2 could significantly reduce thiourea

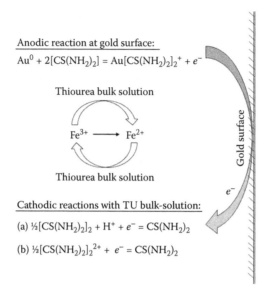

Anodic reaction at gold surface:
$$Au^0 + 2[CS(NH_2)_2] = Au[CS(NH_2)_2]_2^+ + e^-$$

Thiourea bulk solution

$$Fe^{3+} \longrightarrow Fe^{2+}$$

Thiourea bulk solution

Cathodic reactions with TU bulk-solution:

(a) $\frac{1}{2}[CS(NH_2)_2]_2 + H^+ + e^- = CS(NH_2)_2$

(b) $\frac{1}{2}[CS(NH_2)_2]_2^{2+} + e^- = CS(NH_2)_2$

Gold surface

e^-

FIGURE 5.4
Electrochemical reaction mechanism of thiourea leaching of gold.

consumption from 34.4 to 0.57 kg/t, while decreasing leaching time and yielding substantially more gold and silver by preventing the formation of a passivation layer of elemental sulphur. Hence, faster kinetics can be achieved at a low thiourea concentration.

5.2.5 Effect of Pretreatment

For processing the complex ores/concentrates by leaching in thiourea solution, several pretreatments have been suggested for breaking the physical encapsulation and chemical interference to liberate the gold into the solution.

Heat treatment: Pr›eat treatment of gold bearing sulphide ores appears to be promising in comparison to direct leaching (without any pretreatment) (Moussoulos et al., 1984). Difficulty in maintaining the operating conditions of solutions Eh-pH with high reagent consumption is a problem that exists with direct leaching in acidic thiourea solution. In such case, a dosage of more than 5 kg/t thiourea in lixiviant is also not sufficient. Using sulphation roasting of the sulphide ores, followed by leaching of base metals in mild acid or in water, can be beneficial to control the operating condition and lixiviant consumption prior to leaching gold in thiourea solution. An increase in gold recovery (>98%) and decrease in thiourea consumption (~3 kg/t) can be obtained using the pretreatment process.

Pressure oxidation: Pretreatment by pressure oxidation has been found to be useful in some cases and studied by Bruckard et al. (1993) to process the flotation middling of lead-zinc ores via thiourea leaching. Similarly, the bⱥviour of gold-thiourea leaching after pressure oxidation of a refractory ore has been studied to examine the parametric effect of pretreatment temperature, retention time, thiourea concentration, ferric ion concentration, and SO_2 addition. The most important finding of the study indicated the characteristic of pretreatment made thiourea leaching suitable for gold extraction and as economically viable as cyanidation.

Bio-oxidation: Peroxidation of the gold bearing sulphide ores using microbial activity has been investigated by various researchers from time-to-time (Murthy, 1990; Kenna and Moritz, 1991; Deng and Liao, 2002). Such a pre-(bio)oxidation significantly improves the leaching kinetics of gold in thiourea solution, yielding >98% efficiency at the optimal leaching conditions (as discussed in previous sections) from the pyrite and arsenopyrite ores. Nevertheless, the gold bearing carbonaceous ores do not show satisfactory improvement using bio-oxidation, due to the preg-robbing effect.

5.3 Recovery of Gold from Thiourea Leach Liquor

5.3.1 Cementation

Reduction precipitation, or cementation of gold from thiourea leach liquor, is one of the prominent ways to recover the precious metal. Mostly, aluminium, iron, and lead are used for this purpose. However, reports reveal that using Al as the precipitant metal does not yield full recovery, leaving a 2% gold ion in the solution (van Lierde et al., 1982). However, the U.S. Bureau of Mines has reported ~99% recovery of precious metals from thiourea leach liquor while consuming 6.4 kg Al for each kg of the precious metals (gold and silver). In principle, thiourea leach liquor contains iron, and hence, iron is preferred to be used for cementing gold by the following reaction (Groenewald, 1976; Zouboulis et al., 1993; Lee et al., 1997):

$$2Au[SC(NH_2)_2]_2^+ + Fe^0 = 2Au^0 + Fe^{2+} + 4SC(NH_2)_2 \qquad (5.7)$$

In comparison to Fe, the Chinese Institute of Nonferrous Metallurgical Industry reported that using Pb powder for gold cementation yielded better results from HCl-thiourea solution (Wen, 1982). The cementation reaction with lead powder can be written as (Tataru, 1968):

$$\frac{n}{2}Au[SC(NH_2)_2]_2^+ \cdot Cl^- + Pb^0 = \frac{n}{2}Au^0 + Pb[SC(NH_2)_2]_n Cl_2 + \left(\frac{n}{2} - 1\right)Cl_2 \qquad (5.8)$$

A comparatively lower recovery by Fe is described in the study by Wang et al. (2011) that the presence of oxidants in leach liquor negatively affects the gold cementation reaction. The ferric ion significantly hinders the cementation process because of an increased redox potential of the solutions, which signifies the presence of a suitable electron acceptor. It may cause consumption of the Fe powder with ferric ions, as follows:

$$Fe^{3+} + Fe^0 = 3Fe^{2+} \qquad (5.9)$$

The removal of iron powder lowers the availability of this reductant for the cementation of gold, which results in a negative effect on gold cementation. The addition of trisodium citrate into the system can potentially control the redox potential of the solution and form the ferric-citrate complex; however, it can be useful to enhance the gold recovery by controlling the ratio of Fe^{3+}/Fe^{2+}.

5.3.2 Adsorption onto Activated Carbon

Gold recovery from a thiourea leach liquor was primarily studied by Soviet scientists (Lodeishchikov et al., 1968) and has been practiced at New England Antimony Mines (New South Wales). High carbon loading (6–8 kg/t carbon) with a satisfactory efficiency of 90% gold from a leach liquor of 27 mg/L Au by charging 20 g/L carbon-in-pulp has been achieved, but with a high co-adsorption loss (up to 30%) of thiourea (Schulze, 1984). A partial thiourea can be recovered back by hot water back-washing of the adsorbed carbon. The adsorption process follows the Langmuir monolayer isotherm over a wide range of gold concentrations (Zhang et al., 2004). The equilibrium loading of gold decreases with increasing thiourea concentration, pH, and temperature. The presence of Fe^{3+} and Fe^{2+} ions at a low level (up to 700 mg/L) does not affect the equilibrium loading; however, a Cu^{2+} ion significantly reduces it, presumably by competitive adsorption. In the initial stage, the rate of gold adsorption approximates first-order kinetics and is controlled by the diffusion of the gold-thiourea complex to the carbon surface. A change in thiourea concentration does not affect the initial rate of adsorption, nor does the presence of 5 g/L Fe^{2+} ions, but an increase of initial gold concentration, agitation, and temperature greatly affects the adsorption rate. A high concentration of Fe^{3+} ion substantially suppresses the adsorption rate constant, probably due to the deleterious oxidation products of thiourea. Ag^+ and Cu^{2+} ions also strongly reduce the rate by competitive adsorption. Figure 5.5 depicts a detailed flow sheet of thiourea leaching and recovery of gold.

5.3.3 Adsorption by Resin Ion-Exchange

For adsorbing the cationic gold complex, $Au[SC(NH_2)_2]_2^+$ use of strong cation exchanger resins is as effective as activated carbon or charcoal (Groenewald, 1977). As the cationic gold speciation in thiourea predominantly exists in

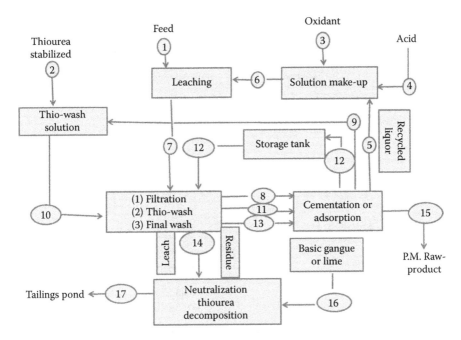

FIGURE 5.5
A detailed flow sheet of thiourea leaching and recovery of gold.

acidic range (<4 pH), the cation exchange capacity of resins significantly changes with respect to the pH and gold concentration within the solution. An alkaline sodium thiosulfate solution and a solution of bromine in hydrochloric acid have been reported to be the most promising eluent solutions; however, the first is preferable. Figure 5.6 indicated the recovery of gold and silver from acidic thiourea leach solution by the ion-exchange resin.

5.3.4 Electrowinning

The electrolytic reduction of gold from $Au[SC(NH_2)_2]_2^+$ complex solution can be achieved by the following reaction:

$$Au[SC(NH_2)_2]_2^+ + e^- = Au + 2SC(NH_2)_2 \quad E = -0.38 \text{ V} \quad (5.10)$$

With a cathodic potential range of −0.15 to −0.38 V, it follows a diffusion-controlled reaction (Groenewald, 1977). However, as such thiourea does not attribute to cathodic reduction of gold; it is the formamidine disulphide, an oxidation product of thiourea, that is supposed to reduce on the cathodic surface. Notably, a separate anodic and cathodic compartment is needed to prevent the formation of decomposition products of thiourea at the anodic compartment, which may redissolve the deposited gold from the

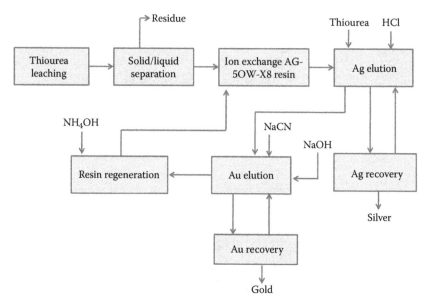

FIGURE 5.6
Flow sheet of gold recovery from acidic thiourea leach solution by ion-exchange resin.

cathode surface. Use of a Pb-anode minimizes the oxidation decomposition of thiourea, and applying a low current density with high catholyte circulation rate yields a better current efficacy for gold deposition.

5.4 Applications of Gold-Thiourea Leaching

Several applications of thiourea leaching of gold bearing ores have been demonstrated on the laboratory and pilot scale after a fine grinding (Kušnierová et al., 1993), mechano-chemical milling (Baláž et al., 2003), sulfation roasting (Moussoulos et al., 1984; Bilston et al., 1990), microbial oxidation (Caldeira and Ciminelli, 1993; Deng and Liao, 2002), and pressure oxidation (Yen and Wyslouzil, 1986; Bruckard et al., 1993). The pretreated ore can be directly leached in thiourea solution without the subsequent step of neutralization, which is always required in the cyanidation process. The most prominent application is on a high content of cyanicides (such as antimony or sulphide) ores by pressure leaching or leaching after a bio-oxidation. An industrial application for processing a gold–antimony concentrate was demonstrated in Australia at the New England Antimony Mine (Hisshion and Waller, 1984); however, the leaching efficiency has not always been as prominent as with cyanidation.

Bio-oxidation followed by acidic thiourea leaching on a 450 t pilot scale (heap leaching) has been conducted by Newmont at the Carlin mine. At a

30–45 L/min flow rate onto the heap, pH < 2.5, redox potential 0.43–0.5 V, gold leaching in thiourea solution (concentration 10 g/L) yielded a poor leaching; only 29% over 110 days. Recovery from the leach liquor by adsorption onto activated carbon or cation-exchange resin was ineffective. The recirculation of thiourea solution formed the elemental sulphur, which got surface coated onto the carbon/resin, and reduced the adsorption capacity.

5.5 Environmental Impact, Limitations, and Challenges for Gold-Thiourea Leaching

While the toxicity of thiourea is far less than cyanide (a lethal dosage of thiourea is 10 g/kg), it is not a totally environmentally friendly reagent. Actually, it is a debatable topic because thiourea is used in thyroid treatment for humans as a non-carcinogenic reagent (IARC, 1974; Shubik, 1975). The various degradation products of thiourea, like urea, ammonia, formamidine disulphide, sulphur, cyanamide, carbon dioxide, sulphate, and nitrate ions, are problematic when discharged and need a proper effluent treatment (Preisler and Berger, 1947; Gupta, 1963; Hiskey and DeVries, 1992). The cyanamide and formamidine disulphide are unstable and short-lived. The half-life of thiourea decomposition in surface water and soil is up to 168 h under aqueous aerobic biodegradation and up to 336 h in groundwater without biodegradation (Howard, 1991). Heavy metals dissolution in thiourea from minerals/soils can cause soil and water pollution.

It must be noted that in the absence of any oxidizing agent, thiourea alone in lixiviant is less effective. $Fe_2(SO_4)_3$ is preferred as a cheaper reagent than H_2O_2, but it also increases thiourea consumption. An excess of $Fe_2(SO_4)_3$ suppresses the leaching of gold. An initial concentration of thiourea and $Fe_2(SO_4)_3$ along with the rate of oxidation of thiourea are the most influential factors affecting the leaching of gold. Formation of degradation products in successive oxidation stages of thiourea makes it difficult to control the leaching system. High consumptions of reagents including $SC(NH_2)_2$, $Fe_2(SO_4)_3$, and H_2SO_4 make the process costlier than using cyanidation. Therefore, without carefully handling the complexity, the implementation of thiourea leaching in gold extraction is quite a difficult task.

References

Baláž, P., Ficeriová, J., Leon, C.V. 2003. Silver leaching from a mechanochemically pretreated complex sulphide concentrate. *Hydrometallurgy.* 70(1): 113–119.

Bilston, D., Bruckard, W., McCallum, D., Sparrow, G., Woodcock, J. 1990. Comparison of methods of gold and silver extraction from hellyer pyrite and lead-zinc flotation middlings. In: Gray, P.M.J., Bowyer, G.J., Castle, J.F., Vaughan, D.J., Warner, N.A. (Eds), *Sulphide Deposits—Their Origin and Processing*. Springer, Dordrecht, pp. 207–221.

Brent Hiskey, J., Atluri, V. 1988. Dissolution chemistry of gold and silver in different lixiviants. *Mineral Procesing and Extractive Metallurgy Review*. 4(1–2): 95–134.

Bruckard, W., Sparrow, G., Woodcock, J. 1993. Gold and silver extraction from hellyer lead-zinc flotation middlings using pressure oxidation and thiourea leaching. *Hydrometallurgy*. 33(1–2): 17–41.

Caldeira, C., Ciminelli, S. 1993. Thiourea leaching of a refractory gold ore. In: *XVIII International Mineral Processing Congress*, Aus.I.M.M., Melbourne, VIC, pp. 1123–1128.

Celik, H. 2004. Extraction of gold and silver from a turkish gold ore through thiourea leaching. *Minerals and Metallurgical Processing*. 21(3): 144–148.

Deng, T., Liao, M. 2002. Gold recovery enhancement from a refractory flotation concentrate by sequential bioleaching and thiourea leach. *Hydrometallurgy*. 63(3): 249–255.

Deschenes, G., Ghali, E. 1988. Leaching of gold from a chalcopyrite concentrate by thiourea. *Hydrometallurgy*. 20(2): 179–202.

Gönen, N. 2003. Leaching of finely disseminated gold ore with cyanide and thiourea solutions. *Hydrometallurgy*. 69(1): 169–176.

Groenewald, T. 1976. The dissolution of gold in acidic solutions of thiourea. *Hydrometallurgy*. 1(3): 277–290.

Groenewald, T. 1977. Potential applications of thiourea in the processing of gold. *Journal of the Southern African Institute of Mining and Metallurgy*. 77(11): 217–223.

Gupta, P.C. 1963. Analytical chemistry of thiocarbamides. *Fresenius' Journal of Analytical Chemistry*. 196(6): 412–431.

Habashi, F. 1969. *Extractive Metallurgy*, vol. 1, General principles, Gordon and Breach. Science Publishers. Inc., New York.

Hiskey, J.B., DeVries, F.W., 1992. Environmental considerations for alternates to cyanide processing. In: Chander, S., Richardson, P.E., El-Sahll, H. (Eds), *Emerging Process Technologies for a Cleaner Environment*. AIME, Phoenix, AZ, pp: 73–80.

Hiskey, J., DeVries, F. 1992. Emerging process technol. In: *Cleaner Environ. Proc. Symp*, pp. 73–80.

Hisshion, R., Waller, C. 1984. Recovering gold with thiourea. *Minining Magazine*. 151(3): 237.

Howard, P.H. 1991. *Handbook of Environmental Degradation Rates*. CRC Press, Boca Raton, FL.

Huyhua, J., Zegarra, C., Gundiler, I. 1989. A comparative study of oxidants on gold and silver dissolution in acidic thiourea solutions. In: Jha, M.C., Hill, S.D. (Eds), *Precious Metals '89*, The Minerals, Metals and Materials Society, Pittsburgh, PA, pp. 287–303.

Ilyas, S., Lee, J.C. 2014. Biometallurgical recovery of metals from waste electrical and electronic equipment: A review. *ChemBioEng Reviews*. 1(4): 148–169.

Kenna, C., Moritz, P. 1991. The extraction of gold from bioleached pyrite using novel thiourea leaching technology. In: *Proceedings of World Gold'91, Gold Forum on Technology a Practice, Second Australia IMM-SME Joint Conference*. Cairns, QLD, pp. 21–26.

Kušnierová, M., Šepelák, V., Briančin, J. 1993. Effects of biodegradation and mechanical activation on gold recovery by thiourea leaching. *JOM Journal of the Minerals, Metals and Materials Society*. 45(12): 54–56.

Lee, H.K., Kim, S.G., Oh, J.K. 1997. Cementation bᵣaviour of gold and silver onto Zn, Al and Fe powders from acid thiourea solutions. *Canadian Metallurgical Quarterly*. 36: 149–155.

Li, J., Miller, J.D., 1997. Thiourea decomposition by ferric sulfate oxidation in gold-leaching systems. *SME Annual Meeting*, SME, Denver, CO, Preprint no. 97–146.

Li, J., Miller, J. 2002. Reaction kinetics for gold dissolution in acid thiourea solution using formamidine disulphide as oxidant. *Hydrometallurgy*. 63(3): 215–223.

Li, J., Miller, J.D. 2006. A review of gold leaching in acid thiourea solutions. *Mineral Processing and Extractive Metallurgy Review*. 27(3): 177–214.

Lodeishchikov, V., Panchenko, A., Shamis, L. 1972. Use of thiourea for extracting au from cu-au containing ores and concentrates. *Nauchn Trudy, Irkutsk. N-I Inst. Redk. Tsvet. Metallov*. 27: 100–108.

Lodeishchikov, V., Panchenkov, A., Briantseva, L. 1968. Use of thiourea as a solvent in the extraction of gold from an ore. *Nauoh. Tr., Irkutsk. Gos. Nauoh.-Issled. Inst. Redk. Tsvet. Metal*. 19: 72–84.

McInnes, C.M., Sparrow, G., Woodcock, J. 1989. Thiourea leaching of gold from an oxidized gold-copper ore. In: *Gold Forum on Technology and Practice-World Gold 89*. Society for Mining, Metallurgy and Exploration Inc., Littleton, CO, pp. 305–313.

Moussoulos, L., Potamianos, N., Kontopoulos, A. 1983. Recovery of gold and silver from arseniferous pyrite cinders by acidic thiourea leaching. In: Kudryk, V., Corrigan, D.A., Liang, W.W. (Eds), *Precious Metals: Mining, Extraction and Processing*. TMS-AIME, Warrendale, PA, pp. 323–335.

Munoz, G., Miller, J. 2000. Noncyanide leaching of an auriferous pyrite ore from Ecuador. *Minerals and Metallurgical Processing*. 17(3): 198–204.

Murthy, D. 1990. Microbially enhanced thiourea leaching of gold and silver from lead-zinc sulphide flotation tailings. *Hydrometallurgy*. 25(1): 51–60.

Örgül, S., Atalay, U. 2000. Gold extraction from kaymaz gold ore by thiourea leaching. *Developments in Mineral Processing*. 13: C6-22–C26-28.

Örgül, S., Atalay, Ü. 2002. Reaction chemistry of gold leaching in thiourea solution for a Turkish gold ore. *Hydrometallurgy*. 67: 71–77.

Plaskin, I., Kozhukhova, M. 1941. The solubility of gold and silver in thiourea. *Doklady Akademii Nauk SSSR*. 31: 671–674.

Plaskin, I.N., Kozhukhova, M. 1960. Dissolution of gold and silver in solutions of thiourea. *Sbornik Nauchnyhk Trudov, Institut Tsvetnykh Metallov*. 33: 107–119.

Preisler, P.W., Berger, L. 1947. Oxidation-reduction potentials of thiol-dithio systems: Thiourea-formamidine disulphide 1. *Journal of the American Chemical Society*. 69(2): 322–325.

Pyper, R., Hendrix, J. 1981. Extraction of gold from a carlin-type ore using thiourea. *Interfacing Technologies in Solution Mining*. 93–108. J-GLOBAL ID: 200902005002625842.

Schulze, R.G. 1984. New aspects in thiourea leaching of precious metals. *JOM Journal of the Minerals, Metals and Materials Society*. 36(6): 62–65.

Shubik, P. 1975. Potential carcinogenicity of food additives and contaminants. *Cancer Research*. 35(11 Part 2): 3475–3480.

Songina, O., Ospanov, K.K., Muldagalieva, I.K., Sal'nikov, S. 1971. Dissolution of gold with the use of thiourea in a hydrochloric acid medium. *Izvestiya Akademii Nauk Kazakhskoi SSR, Seriya Khimicheskaya*. 21: 9–11.

Sparrow, G.J., Woodcock, J.T. 1995. Cyanide and other lixiviant leaching systems for gold with some practical applications. *Mineral Processing and Extractive Metallurgy Review*. 14(3–4): 193–247.

Tataru, S. 1968. Precipitation par cementation de l'or en solutions acid. *Revue Roumaine de Chimie*. 1043–1049.

Tremblay, L., Deschenes, G., Ghali, E., McMullen, J., Lanouette, M. 1996. Gold recovery from a sulphide bearing gold ore by percolation leaching with thiourea. *International Journal of Mineral Processing*. 48(3–4): 225–244.

Van Lierde, A., Ollivier, P., Leosille, M. 1982. Développement du nouveau procedé de traitment pour le mineraux de salsigne. *Ind. Min. Les* Tech. 1a. 399–410.

Van Staden, P. 1989. In-stope leaching with thiourea. *Journal of the Southern African Institute of Mining and Metallurgy*. 89(8): 221–229.

Wang, Z., Li, Y., Ye, C. 2011. The effect of tri-sodium citrate on the cementation of gold from ferric/thiourea solutions. *Hydrometallurgy*. 110(1): 128–132.

Wen, C.D. 1982. Studies and prospects of gold extraction from carbon bearing clayey gold ore by the thiourea. In: *Proceedings XIV International Mineral Processing Congress*. Toronto, ON, October 17–23, 1982.

Yen, W.T., Wyslouzil, D.M. 1986. Pressure oxidation and thiourea extraction of refractory gold ore. In: *Gold 100, Proceedings of the International Conference on Gold*, Vol. 100, SAIMM, Johannesburg, South Africa, pp. 579–589.

Zhang, H., Ritchie, I.M., La Brooy, S.R. 2004. The adsorption of gold thiourea complex onto activated carbon. *Hydrometallurgy*. 72(3): 291–301.

Zhang, Y., Liu, S., Xie, H., Zeng, X., Li, J. 2012. Current status on leaching precious metals from waste printed circuit boards. *Procedia Environmental Sciences*. 16: 560–568.

Zouboulis, A., Kydros, K., Matis, K. 1993. Recovery of gold from thiourea solutions by flotation. *Hydrometallurgy*. 34(1): 79–90.

6

Halide Leaching of Gold

Sadia Ilyas*, Humma Akram Cheema*, and Jae-chun Lee†

6.1 Chemistry for Halide Leaching of Gold

The two common forms of gold in +1 and +3 states are B-type metal ions; therefore, their complexation stability decreases as the electronegativity of the donor ligand increases, forming the stability order of $I^- > Br^- > Cl^- > F^-$. Using the Hard and Soft Acid-Base (HSAB) theory, the complexation of Au^{3+} with halides, hard donors will be more stable. It indicates that complexes type AuL_4^{3+} with soft ligands will easily reduce to a +1 state (John, 1972).

All known AuL_4^{3+} complexes have the electronic configuration $4f^{145}5d^8$, and exhibit low spin diamagnetic properties. Compounds like $AuCl_3$ form dimmers to fulfil the requirement of four coordination numbers by Au^{3+}. Chloride leaching among the halides has been widely employed prior to the cyanidation process and is discussed here. The chlorine generation is mainly carried out in two ways: (i) by reactions of sodium hypochlorite with HCl, and (ii) production of anodic chlorine via electro-dissociation of concentrated HCl at cathodic compartment. The reaction with sodium hypochlorite for chlorine takes place as:

$$NaOCl + NaCl + 2HCl = 2NaCl + Cl_{2(g)} + H_2O \qquad (6.1)$$

The chlorine generation in an electrolytic cell at the anodic compartment can be obtained as the reactions:

$$At\ anode: 2Cl^- = Cl_{2(g)} + 2e^- \qquad (6.2)$$

$$At\ cathode: 2H^+ + 2e^- = H_{2(g)} \qquad (6.3)$$

---
* Mineral and Material Chemistry Lab, Department of Chemistry, University of Agriculture Faisalabad, Pakistan.
† Minerals Resources Research Division, Korea Institute of Geoscience and Mineral Resources, Daejeon, South Korea.

The gaseous chlorine has high solubility in acidic water as (Snoeyink and Jenkins, 1979; Lee and Srivastava, 2016):

$$Cl_{2(g)} = Cl_{2(aq)} \tag{6.4}$$

At a pH above 2.5, the aqueous chlorine predominantly forms HOCl as (Marsden and House, 2006):

$$Cl_{2(aq)} + H_2O = HCl + HOCl \tag{6.5}$$

Both acids, HCl and HOCl, completely dissociate in aqueous solutions as:

$$HCl = H^+ + Cl^- \tag{6.6}$$

$$HOCl = H^+ + OCl^- \tag{6.7}$$

Further, the soluble chlorine can react with the chloride ions to form the trichloride ions under high acidic condition (< pH 3):

$$Cl_{2(aq)} + Cl^- = Cl_3^- \tag{6.8}$$

6.2 Development History of Gold-Halide Leaching

Among the halogens, chlorine was first applied in the 1800s, far before cyanidation was introduced. It was found to be more effective for processing gold bearing sulphide ores and difficult in amenability to gravity concentration and amalgamation. Later in 1846, bromine was introduced as a solvent for gold whose leaching kinetics are greatly enhanced in the presence of a protonic cation and an oxidizing agent (Filmer et al., 1984; Kalocsai, 1984; von Michaelis, 1987). Although the cyanidation of gold diminished the potential of halogen in general, the recent interest in halide/s leaching came during the 1990s (Tran et al., 2001). Gold leaching in aqua regia is also an example of chlorination leaching in which an enhanced kinetics of gold leaching is provided by the strong oxidizing environment of NOCl, forming a soluble Au^{3+} species as in the following reactions (Lee and Srivastava, 2016):

$$3HCl + HNO_3 = NOCl + Cl_2 + 2H_2O \tag{6.9}$$

$$Au + NO_3^- + 4H^+ = Au^{3+} + 2H_2O + NO \tag{6.10}$$

$$Au^{3+} + 4Cl^- = AuCl_4^- \tag{6.11}$$

$$Au + HNO_3 + 4HCl = HAuCl_4 + 2H_2O + NO \tag{6.12}$$

The stability of gold-halide complexes is dependent on the Eh-pH in the solution, composition (with respect to halide concentration), and the nature of ores to be processed (Sergent et al., 1992; Tran et al., 2001). A residual quantity of oxidant must be maintained to keep a high Eh in the solution, avoiding precipitation of metallic gold (Tran et al., 2001). Overall, the stability of halides is in the order of $I^- > Br^- > Cl^- > F^-$, whereas the rate of reaction is $F^- > Cl^- > Br^- > I^-$. Typical conditions used for leaching gold by halogens and their thermodynamic data are listed in Tables 6.1 and 6.2, respectively.

6.3 Chlorine Leaching

The elimination of *in* situ loss of gold via adsorption with carbonaceous matter and the detoxification of free cyanide are the main advantages for using chlorine leaching (Marsden and House, 2006). With dissolved chlorine in an aqueous solution, gold forms chloride complexes as (Radulescu et al., 2008):

$$Au + Cl^- + \frac{3}{2}(Cl_2)_{aq} = AuCl_4^- \tag{6.13}$$

TABLE 6.1

Process Conditions for Gold Halide Leaching and Thermodynamic Data

Process Conditions for Halide Leaching of Gold

Ligand	Oxidant	Processs Conditions	Reagent
Cl^-	$Cl_2/HClO$	5–10g/L Cl_2 or NaCl	Chlorine
Br^-	Br_2	2–5 g/L Br_2, 0–10 g/L NaBr	Bromine
I^-	I_2	1 g/L I_2, 9 g/L NaI	Iodine

TABLE 6.2

Thermodynamic Data of Gold Halide Complexes

Compound	Color	Oxidation State	Partial Pressure (/gm³)	ΔH (kJ/mol)	ΔG (kJ/mol)
AuF	–	+1	–	–75	–
AuCl	Yellow-white	+1	7.8	–35	–16
AuBr	Light yellow	+1	7.9	–19	–15
AuI	Lemon yellow	+1	8.25	1.7	–3
AuF_3	Golden yellow	+3	6.7	–360	–
$AuCl_3$	Red	+3	4.3	–121	–54
$AuBr_3$	Dark brown	+3	–	–67.3	–36
AuF_5	Dark red	+5	–	–	–

$$Au + \frac{3}{2}HOCl + \frac{3}{2}H^+ + \frac{5}{2}Cl^- = AuCl_4^- + \frac{3}{2}H_2O \tag{6.14}$$

The formation of soluble species (at <6 pH) is also in line with the thermodynamic data plotted in Figure 6.1, for the Eh-pH diagram of $Au-Cl_2-H_2O$ system. Interestingly, a direct formation of auric chloride ($AuCl_4^-$) with $Cl_{2(aq)}$ and HOCl differs with the reaction of Cl^- ions due to less oxidative power. In stepwise leaching of gold in the presence of Cl^- ions, an intermediate aurous complex on the surface forms as (Nicol, 1980; Nicol et al., 1987):

$$2Au + 2Cl^- = 2AuCl^- \tag{6.15}$$

Using the HSAB principle again, the high electronegative Cl^- ions (a hard donor atom) tend to form a secondary intermediate compound of high valance (Finkelstein and Hancock, 1974) as with the following reaction in the second step:

$$2AuCl^- = AuCl_2^- + Au \tag{6.16}$$

In a subsequent step, $AuCl_2^-$ either oxidizes to form a stable auric complex (Nicol, 1980; Nicol et al., 1987) or diffuses into a solution as the following reaction, forming the $AuCl_2^-$:

$$AuCl_2^- + 2Cl^- = AuCl_4^- \tag{6.17}$$

The copper that is often associated in gold bearing ores can also get leached in a chlorine environment. By the following reactions for copper in the system, the electro-chemical nature of the chlorine leaching of gold can be described as in Figure 6.2 (Kim et al., 2010):

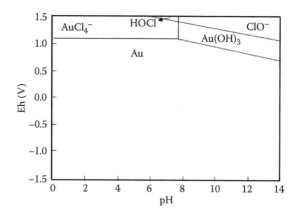

FIGURE 6.1
The Eh-pH diagram of gold complexes in chloride media.

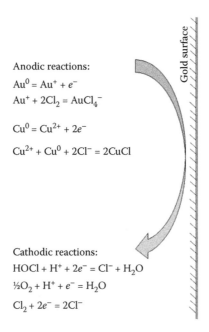

Anodic reactions:

$Au^0 = Au^+ + e^-$

$Au^+ + 2Cl_2 = AuCl_4^-$

$Cu^0 = Cu^{2+} + 2e^-$

$Cu^{2+} + Cu^0 + 2Cl^- = 2CuCl$

Cathodic reactions:

$HOCl + H^+ + 2e^- = Cl^- + H_2O$

$\frac{1}{2}O_2 + H^+ + e^- = H_2O$

$Cl_2 + 2e^- = 2Cl^-$

Gold surface

FIGURE 6.2
Electrochemical mechanism of chloride leaching of gold.

$$Cu + Cl_{2(aq)} = Cu^{2+} + 2Cl^- \qquad (6.18)$$

$$Cu^{2+} + Cu + 2Cl^- = 2CuCl \qquad (6.19)$$

By looking at the critical presence of copper in gold bearing ores, the formation of dissolved species of both metals can easily be depicted from the Eh–pH diagram of Au–Cu–Cl$_2$–H$_2$O system (Figure 6.1).

6.3.1 Effect of Acid Concentration

The fundamentals of electro-chemical kinetics for leaching gold in a chloride solution with the dissolution chemistry have already been elaborated (Finkelstein, 1972; Nicol, 1976; Avraamides, 1982; Yen et al., 1990; Tran et al., 1992a,b; Lee and Srivastava, 2016). A faster weight loss of gold in various mixture solutions of chloride–hypochlorite than the cyanidation process under similar parametric conditions has already been achieved (Tran et al., 2001). The formation of a stable species, AuCl$_4^-$ strongly depends on the solution pH (<3.0; as evident from the Eh–pH diagram in Figure 6.1), with high chloride/chlorine levels (>100 g/L Cl$^-$), elevated temperature, and high low particle size of the processing ore. The dissolved gold complex can reprecipitate by contact with a reductant like sulfidic materials; therefore, the application of the chloride–chlorine systems is limited to processing the oxidized minerals or ores. Notably, the solubility of chlorine increased with respect to increasing acid concentration and forming various soluble species, such as aqueous Cl_2 and Cl_3^- (Lee and Srivastava, 2016). Creating a highly oxidative environment can be helpful to process the ores other than oxidized bodies. Additionally, the redox potential of a leaching system also needs to maintain a speed above 400 mV for a faster kinetics and higher leaching yield of gold.

6.3.2 Effect of Temperature

Nevertheless, the applicability of temperature can enhance the rate of reaction and thus obtain a higher leaching yield (Nicol, 1980; Viñals et al., 1995), but a chloride–chlorine system is quite different than leaching of metals in other mineral acids. The solubility of NaCl increases from 359 to 385 g/L with respect to increasing the temperature from 20°C to 90°C; whereas, the solubility of gaseous chlorine to form the aqueous Cl_2 and Cl_3^- has been found to decline when elevating the temperature ≥60°C (Lee and Srivastava, 2016). In that case, the leaching kinetics would not only be affected by the chemical reactivity but also by the mass transfer phenomena (Kim et al., 2016).

6.3.3 Effect of Initial Cl^- Concentration

The leaching kinetics of gold in a chloride medium is proportionally dependent on the chlorine-chloride concentrations (Nicol, 1980). The leaching efficiency therefore increases with the increasing of the initial concentration of chloride and soluble chlorine in lixiviant solution with an enhanced temperature (preferably <60°C). Gold leaching in chloride solution is much faster than the yield obtained in an alkaline cyanide solution. A rate of 0.008 g/m^2/s gold leaching in cyanide solution has been found to be much less than the leaching rate of 0.3 g/m^2/s obtained in chloride solution by Putnam (1944). A high solubility of chlorine in water compared with oxygen (used in cyanidation) is a plausible reason for this. The presence of 3% NaCl in a chlorine solution has shown a significant effect on gold leaching and may be due to the retarding effect of Cl^- ions on chlorine dissolution (Chao, 1968). Additionally, the amount of initial chlorine in the solution increases the kinetics of gold leaching by shifting the reaction mechanism from diffusion control to a chemically controlled reaction (Lee and Srivastava, 2016).

Other than the previous parametric effects, the effect of roasting on chlorine leaching of a gold bearing refractory concentrate has been studied (Birloaga et al., 2013). An increase in the roasting temperature has been found to be advantageous to improve the removal efficiency of Hg (~94%). The sulphur removal performed by the roasting step could significantly reduce the chlorine consumption and yield a far better leaching of gold (~93%) than that of using only cyanidation (27%).

6.4 Bromine Leaching

Bromine, as a lixiviant for gold was first described in 1846 (von Michaelis, 1987). Liquid bromine exothermally reacts with gold to form Au_2Br_6 of dimeric structure as with Au_2Cl_6. It forms $[LAuBr_3]^-$type complex with nitrogen-donor ligands. The monobromide of gold, like its corresponding monochloride, is sensitive to moisture decomposing to auri- and aurobromide complexes, such as $[LAuBr]^-$ and $[AuBr_2]^-$.

The leaching kinetics of gold with bromine is greatly accelerated in the presence of a protonic cation (e.g., NH_4^+) and an oxidizing reagent (Kalocsai, 1984). A fast leaching of gold is possible at nearly neutral pH conditions; whereas, with the requirement of high redox potential ($E^0 = 0.97V$ vs. SHE) compared with the gold cyanidation ($-0.57V$ vs. SHE), stabilization of the gold bromide complex can be achieved by the addition of a strong oxidant like bromine. Bromine can be introduced to the slurry (as bromide) with hypochlorite or chlorine used as oxidants to convert the bromide to bromine as follows:

$$2Br^- + Cl_2 = 2Cl^- + Br_2 \tag{6.20}$$

$$2Br^- + ClO^- + 2H^+ = Br_2 + Cl^- + H_2O \tag{6.21}$$

After completion of the leaching reaction, the unconsumed bromine gets converted to bromide ions. The dissolution of gold was shown to depend on the bromine–bromide ratio and the associated minerals in the ore (Pesic and Sergent, 1992; Van Meersbergen et al., 1993). The presence of base metals (like Cu, Zn, Al, and their sulphates) has no effect on leaching, but Fe^{2+} and Mn^{2+} are oxidized and increase the bromine consumption. Hence, investigations have been carried out in search of alternative oxidants (like Fe^{3+}, H_2O_2, and NaOCl) to bromine for eliminating the associated problems including corrosive reactions and high vapor pressure, with limited success (Trindade et al., 1994; Sparrow and Woodcock, 1995). The organic bromides (N-halo hydantoins such as Geobrom 3400) can be used to reduce the problems that exist with vapor loss.

6.5 Iodine Leaching

Gold-iodide complexes are the most stable halide complexes in aqueous solutions following the order: $I^- > Br^- > Cl^-$ (Nicol et al., 1987). As can be seen in the Eh-pH diagram of $Au-I_2-H_2O$ system shown in Figure 6.3, the stability region of AuI_2^- complex occupies a wider pH range (up to 13) compared with other halides (Brent Hiskey and Atluri, 1988; Baghalha, 2012). In the

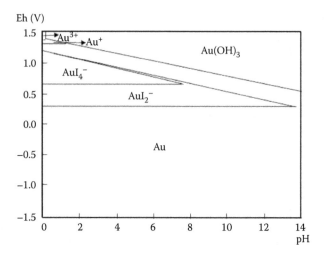

FIGURE 6.3
Eh-pH diagram of gold complexes in iodide media.

presence of iodide, gold oxidizes to form AuI_2^- at redox potential 0.51 V; subsequently going above 0.69 V, the AuI_4^- complex forms whose stability region is greater than the $AuCl_4^-$ and $AuBr_4^-$. Iodine solubilizes in pure water as (Marken, 2006):

$$I_{2(s)} = I_{2(aq)} \tag{6.22}$$

An increase in I^- concentration increases the overall solubility of iodine, forming the tri-iodide at a pH below 9 (Davis et al., 1993):

$$I_{2(aq)} + I^- = I_3^- \tag{6.23}$$

Going above pH 9, the reactions for the formation of a hypoiodite and iodite ion can be written as (Davis et al., 1993):

$$I_3^- + H_2O = IO^- + 2I^- + 2H^+ \tag{6.24}$$

$$3I_3^- + 3H_2O = IO_3^- + 8I^- + 6H^+ \tag{6.25}$$

The presence of tri-iodide ions in an iodide solution above $1.0 \times 10^{-3} M$ (Marken, 2006) acts as an oxidant for the electrochemical mechanism (shown in Figure 6.4) of leaching of gold as follows (Baghalha, 2012):

$$2Au + I_3^- + I^- = 2AuI_2^- \tag{6.26}$$

$$2Au + 3I_3^- = 2AuI_4^- + I^- \tag{6.27}$$

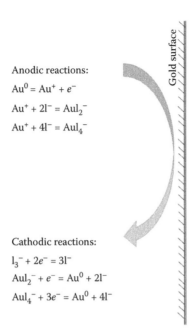

Anodic reactions:

$Au^0 = Au^+ + e^-$

$Au^+ + 2I^- = AuI_2^-$

$Au^+ + 4I^- = AuI_4^-$

Cathodic reactions:

$I_3^- + 2e^- = 3I^-$

$AuI_2^- + e^- = Au^0 + 2I^-$

$AuI_4^- + 3e^- = Au^0 + 4I^-$

FIGURE 6.4
Electrochemical mechanism of iodide leaching of gold.

6.5.1 Effect of pH

As per the wide range of pH for soluble species of gold in iodine solution (Figure 6.3), the leaching of gold has been investigated under a wider pH range than other halides. Gold extraction increases with increasing the pH up to pH 9.0; after that, it decreases rapidly. The predominant formation of oxidized species of iodine, IO_3^- instead of I_3^-, is mainly responsible for this (Davis et al., 1993). Hence, the leaching rate of gold depends on the iodine and iodide concentrations in the solution and is not greatly affected by changes in pH up to 9.0.

6.5.2 Effect of Iodine Concentration

The species AuI_4^- is meta-stable as the iodide ion converted to iodine (Sparrow and Woodcock, 1995) gets reduced as AuI_2^-. The $[I_3^-]/[I^-]$ couple present in the solution can be determined by the Nearst equation, as follows (Brent Hiskey and Atluri, 1998):

$$E = 0.54 - 0.03\log\frac{[I^-]^3}{[I_3^-]} \tag{6.28}$$

In a decreased concentration of iodine, the AuI_2^- and AgI_2^- complexes become unstable and cause a decline in the leaching efficiency of gold. Additionally, the hydroxide precipitation of base metals on gold surfaces hinders the leaching process. The maximum gold leaching was obtained with the iodine/iodide mole ratio at 1:5. The excessive I⁻ under an acidic environment can precipitate gold as AuI; precipitating the residual base metals (Pb_2O as PbI_2 and CuO as CuI) passivates on the surface of gold and hinders the formation of AuI_2^-.

6.5.3 Effect of Oxidant

In the presence of an oxidant, iodide leaching of gold is advantageous in two ways:

  i. By increased recovery of gold, and
  ii. A reduced consumption of iodine.

An addition of 1% H_2O_2 increases the gold leaching up to 95%; however, a further addition of H_2O_2 may precipitate iodine to decline the leaching yield. Hence, the leaching conditions of oxidant/iodide molar ratio, concentration, and pH need to be optimized to avoid the surface passivation by AuI and to maximize the leaching rate. The leaching reaction of gold in an iodide solution in the presence of H_2O_2 can be written as:

$$2Au + 4I^- + H_2O_2 = 2AuI_2^- + 2OH^- \tag{6.29}$$

Alternatively, hypochlorite has been used in which a maximum leaching could be achieved at a [OCl⁻]/[I⁻] molar ratio of 0.25 (Davis and Tran, 1991). I_3^- is formed by reacting hypochlorite with iodide and actively complexes the gold, exhibiting a higher efficiency of gold extraction.

6.6 Applications of Gold-Halide Leaching

A pretreatment of carbonaceous or sulfidic ores (either by roasting or pressure oxidation) is commonly required before undergoing halide leaching to acquire relatively inert ores and less reagent consumption (Sparrow and Woodcock, 1995). A bench scale study carried out at room temperature with 20% pulp density in the chloride solution containing 25 g/L NaOCl and 0.35 M HCl has shown that NaCl concentration increased the leaching yields of gold. Pressure oxidation followed by NaOCl leaching of gold bearing copper concentrate yielded a 90% extraction under these conditions: 25 g/L NaOCl, 200 g/L NaCl, 0.35 M HCl, time 1 h (Puvvada and Murthy, 2000). But the hypochlorite leaching tested by the McDonald Gold Mine oxide ore in Montana could leach only 68% gold as compared with 73% gold with the cyanidation process (McNulty, 2001). Interest in the use of chloride leaching for processing the copper anode slime has continued in several copper refineries (Herreros et al., 1999; Tran et al., 2001). Some of the scaled-up process suitable for industrial application is described in the following paragraphs.

The Minataur™ process: The Minataur™ process, developed by Mint–(South Africa), comprises oxidative leaching of the ore body followed by gold separation using solvent extraction technique prior to precipitating the high-purity gold (Feather et al., 1997). Suitable feeds include silver-refining anode slimes, gold-electrowinning cathode sludge, zinc-precipitation filtrates, gravity-gold concentrates, and the residues from mill liners in gold plants. Leaching performed with a continuous passing of chlorine in 5 M HCl solution efficiently yielded gold in the leach liquor. For recovery purposes, separately extracted gold into organic solvent can be recovered via direct reduction in oxalic acid or sulphur dioxide. In addition, the Gravitaur™ process has been developed that incorporates gravity concentrates as the feed material.

Intec gold process: Primarily used for copper processing, copper sulphide is typical feed to be leached at 85°C using 280 g/L NaCl + 28 g/L NaBr solution. Gold can be easily leached in a solution of Eh 1.2 V, which can be precipitated at the dropped Eh 0.80 V of solution (Moyles, 1999; Severs, 1999).

PLATSOL process: An alternative to cyanide, this process was developed jointly by the University of British Columbia, O'Kane Consultants Ltd., and Lakefield Research (Canada) for processing the flotation sulphide concentrates from the Polymet Mining Company located in Minnesota. The process consists of single leaching of base metals (copper and nickel) along with the precious metals (gold and PGMs). The process generates an oxidizing environment under pressure (at $\geq 200°C$) in the autoclave capable to leach Au^{3+} as chloro-complexes in the presence of a 5–10 g/L NaCl solution. The novelty of this process relies on supplying the limited amount of chloride that is required only to sufficiently leach the gold and PGMs, and preventing the precious metals to their reverese solubility. The leaching reaction of gold can be written as:

$$Au + \frac{1}{4}O_2 + \frac{1}{2}H_2SO_4 + 4Cl^- = AuCl_4^{3-} + \frac{1}{2}SO_4^{2-} + \frac{1}{2}H_2O \qquad (6.30)$$

Ore grinding with ceramic instead of iron balls is required to prevent cementation of gold chloride (Ferron et al., 2000). The concentrate (14.7% Cu, 3.0% Ni, 0.14% Co, 26.7% S, 1.4 g/t Au, 2.2 g/t Pt, and 9.9 g/t Pd) leached under pressure oxidation at 225°C, pulp density 11%, retention time 2 h, pO_2 689 kPa leads to yield 89.4% Au along with 99.6% Cu, 98.9% Ni, 96% Co and Pt, and 94.6% Pd. For recovery purposes, adsorption by carbon-in-pulp does not require a prior neutralization step.

6.7 Recovery of Gold from Halide Leach Liquors

6.7.1 Carbon Adsorption

Adsorption onto activated carbon is a prominent way to recover the gold following the sorption capacity as the order: $AuCl_4^- > Au(CN)_2^- > Au(SCN)_2^- > Au(CS(NH_2)_2)_2^+ > Au(S_2O_3)_2^{3-}$. However, the adsorption mechanism is believed to follow the reduction of chloride complex to metallic form as in the following reactions:

$$AuCl_4^- + 3e^- = Au^0 + 4Cl^- \quad E^0 = 1.00 \text{ V} \tag{6.31}$$

$$AuCl_2^- + e^- = Au^0 + 2Cl^- \quad E^0 = 1.16 \text{ V} \tag{6.32}$$

$$AuCl_4^- + 2e^- = AuCl_2^- + 2Cl^- \quad E^0 = 0.92 \ V \tag{6.33}$$

The requirement of reducing electrons to drive the previous reactions is fulfilled by the activated carbon and by the anodic reaction as follows:

$$C + H_2O = 4H^+ + CO_2 + 4e^- \quad E^0 = 0.21 \text{ V} \tag{6.34}$$

The reduction rate of gold chloride complex at carbon surface is controlled diffusion via boundary layer, hence deposition of gold initially occurs at surface available sites, spreading the spherical gold particles at the carbon surfaces. The deposition rate is independent of pH < 7.0, and decreases remarkably at >7.0 pH. The oxidizing power of NaClO, which is often present in chloride systems, shows adverse effects on the reduction of gold chloride complex to metallic form. Concentration of chloride ions in the gold bearing leach solution is the most important factor to be maintained in the adsorption process. Because of an increase in chloride ion concentration, the E^0 value decreases, which decreases the gap between redox potential of activated carbon and the reduction of gold. This causes a decreased driving force for adsorption, resulting in a low loading capacity.

As the adsorption process is actually a reduction of gold-chloride complex onto carbon surface, the elution is also not desorption of the metal from loaded carbon. Rather, it is a redissolution of gold from its metallic form to a soluble species. It can be achieved by an acid-thiourea or amine-thiosulfate solution. Alternatively, the adsorbed carbon can be burnt at a high temperature to yield directly the metallic gold.

6.7.2 Cementation

Cementation is one of the simple ways to recover gold from the halide leach liquor, especially from the commonly used chloride leaching. Reductive precipitation of gold is performed by contacting the leach liquor with a metal having a potential above that of the precious metal in the electrochemical series. In the case of Au^{3+}, the metals (M, generally standing for iron and aluminium), which can form trivalent ions in chloride solution, are usually employed due to their half-cell reactions as follows:

$$Au^{3+} + 3e^- = Au^0 \quad E^0 = 1.40 \text{ V} \tag{6.35}$$

$$Fe^{3+} + 3e^- = Fe^0 \quad E^0 = -1.21 \text{ V} \tag{6.36}$$

$$Al^{3+} + 3e^- = Al^0 \quad E^0 = -1.66 \text{ V} \tag{6.37}$$

The precipitation reaction of gold can be commonly written as:

$$Au^{3+} + M^0 = Au^0 + M^{3+} \tag{6.38}$$

In principle, the metal M should be dissolved to precipitate the gold from leach liquor, hence the pH is always maintained in acidic range.

6.7.3 Solvent Extraction

Gold dissolves in chloride complex by forming anionic species, and hence, the extraction of gold can be achieved by amine base organic solvents. During extraction of gold-chloride complex, the concentration of free acid is a vital factor that can be co-extracted with amine solvents to adversely affect the loading capacity for gold. Using Hostarex A327 in n-decane, the extraction equilibria is found to be dependent on the ionic strength and of exothermic nature. The extraction reaction can be written as (Martineza et al., 1999):

$$(R_3NH^+Cl^-)_{org} + (AuCl_4^-)_{aq} = (R_3NH^+AuCl_4^-)_{org} + (Cl^-)_{aq} \tag{6.39}$$

The extraction of gold chloride complex can also be extracted using phosphine oxide extractants by following the solvation mechanism (Martinez et al., 1996; Mironov and Natorkhina, 2012). The extraction reaction can be written as:

$$H_{aq}^+ + (AuCl_4^-)_{aq} + nH_2O_{aq} + mTBP_{org} = H^+AuCl_4^- \cdot mTBP \cdot nH_2O \tag{6.40}$$

The extracted gold from loaded organic material in both cases can be easily stripped in a thiourea-HCl solution at mole ratio 1:1 (Narita et al., 2006).

6.8 Environmental Impact, Limitations and Challenges for Halide Leaching of Gold

The corrosiveness and toxicity of halides are difficult from an environmental perspective and need a greater process control on discharge of the final effluent. In gaseous form the chlorine and bromine are dangerous to inhale, causing irritation; high dosage may result in memory loss.

Processing loss of bromide or iodide by adsorption on the gangue minerals or precipitation as insoluble copper/lead compounds require high reagent consumptions for low-grade ore applications. The controlled ORP leaching is difficult to operate, and leaching in a highly oxidative environment in acid solutions dissolves the base metals, presenting great challenges during the separation process. Controlled temperature reactions would be preferential to reduce the operational cost. Chlorine dissolution in aqueous solution has a major role in gold leaching, and the replacement of HCl with cheaper and less aggressive reagents will be required. Iodine has been identified as a good oxidant (<11 pH) and better than HOCl with higher leaching kinetics of gold. Nevertheless, iodine continues to be underemployed as a lixiviant for gold leaching, largely because of cost.

References

Avraamides, J. 1982. Prospects for alternative leaching systems for gold: A review. In: *Proceedings, Symposium on Carbon-in-Pulp Technology for the Extraction of Gold*, Aus.I.M.M., Melbourne, VIC, pp. 369–391.

Baghalha, M. 2012. The leaching kinetics of an oxide gold ore with iodide/iodine solutions. *Hydrometallurgy*. 113–114: 42–50.

Birloaga, I., De Michelis, I., Ferella, F., Buzatu, M., Vegliò, F. 2013. Study on the influence of various factors in the hydrometallurgical processing of waste printed circuit boards for copper and gold recovery. *Waste Management*. 33(4): 935–941.

Brent Hiskey, J., Atluri, V.P. 1988. Dissolution chemistry of gold and silver in different lixiviants. *Mineral Processing and Extractive Metallurgy Review*. 4: 95–134.

Chao, M.S. 1968. The diffusion coefficients of hypochlorite, hypochlorous acid, and chlorine in aqueous media by chronopotentiometry. *Journal of the Electrochemical Society*. 115(11): 1172–1174.

Davis, A., Tran, T. 1991. Gold dissolution in iodide electrolytes. *Hydrometallurgy*. 26(2): 163–177.

Davis, A., Tran, T., Young, D.R. 1993. Solution chemistry of iodide leaching of gold. *Hydrometallurgy*. 32: 143–159.

Feather, A., Sole, K.C., Bryson, L.J. 1997. Gold refining by solvent extraction—The Minataur™ Process. *Journal of the Southern African Institute of Mining Metallaurgy*. 97(4): 169.

Ferron, C.J., Fleming, C.A., Dreisinger, D.B., O'Kane, P.T. 2000. Single-step pressure leaching of base and precious metals (gold and PGMs) using the PLATSOLTM process. Presented at the *ALTA 2000 Nickel/Cobalt Conference*, May 15–18, 2000, Perth, WA.

Filmer, A.O., Lawrence, P.R., Hoffman, W. 1984. A comparison of cyanide, thiourea and chlorine as lixiviants for gold. In: *Regional Conference: Proceedings on Gold Mining, Metallurgy and Geology*, Aus.I.M.M., Melbourne, VIC, October 1984, pp. 1–8.

Finkelstein, J.J. 1972. The goring ox: Some historical perspectives on deodands, for-feitures, wrongful death and the Western notion of sovereignty. *Temple Law Quarterly*, vol. 46, p. 169.

Finkelstein, N.P., Hancock, R.D. 1974. A new approach to the chemistry of gold. *Gold Bulletin.* 7(3): 72–77.

Herreros, S., Qiuiroz, R., Vinal, J. 1999. Dissolution kinetics of copper, white metal and natural chalcocite in chlorine/chloride media. *Hydrometallurgy.* 51: 345–357.

John, A.D. 1972. *Lange's Handbook of Chemistry.* 15th edition. McGraw-Hill Inc., New York. ISBN 0-07-016384-7.

Kalocsai, G.I.Z. 1984. Improvements in or relating to the dissolution of noble metals. Australian Provisional Patent 30281/84.

Kim, E.-Y., Lee, M.-S., Jha, J.-C., Yoo, M.K., Jeong, K. J. 2010. Leaching b›avior of Kim copper using electro-generated chlorine in hydrochloric acid solution. *Hydrometallurgy.* 100: 95–102.

Kim, M.S., Park, S.W., Lee, J.C., Choubey, P.K. 2016. A novel zero emission concept for electrogenerated chloride leaching and its application to extraction of platinum group metals from spent automotive catalyst. *Hydrometallurgy.* 159: 19–27.

Lee, J.C., Srivastava, R.R. 2016. Leaching of gold from spent/end-of-life mobile phone-PCBs. In: Sabir, S. (Ed.), *The Recovery of Gold from Secondary Source.* Imperial College Press, London, UK, pp. 7–56.

Marken, F. 2006. The electrochemistry of halogens. In: Bard, A.J., Stratmann, M., Scholz, F., Pickett, C.J. (Eds), *Encyclopedia of Electrochemistry, Inorganic Chemistry.* Chapter 9, Vol. 7, Wiley-VCH, Weinheim, Germany, pp. 291–297.

Marsden, J.O., House, C.L. 2006. *The Chemistry of Gold Extraction.* 2nd Edition. The Society of Mining, Metallurgy, and Exploration Inc. (SME), Littleton, CO.

Martinez, S., Sastre, A.M., Alguacil, F.J. 1999. Solvent extraction of gold(III) by the chloride salt of the tertiary amine Hostarex A327. Estimation of the interaction coefficient between $AuCl_4^-$ and H^+. *Hydrometallurgy.* 52(1): 63–70.

Martinez, S., Sastre, A., Miralles, N., Alguacil, F.J. 1996. Gold(III) extraction equilib-rium in the system Cyanex 923-HCl-Au(III). *Hydrometallurgy.* 40(1–2): 77–88.

McNulty, T. 2001. Cyanide substitutes. *May Mining Magazine*, pp. 256–260.

Mironov, I.V., Natorkhina, K.I. 2012. On the selection of extractant for the precipita-tion of high-purity gold. *Russian Journal of Inorganic Chemistry.* 57(4): 610–615.

Moyles, J. 1999. The Intec Copper Process demonstration plant operating experi-ence and results from the 1999 campaign. Copper Concentrate Treatment Short Course, unpublished presentation given at Copper 99, Phoenix, AZ.

Van Meersbergen, M.T., Lorenzen, L., van Deventer, J.S.J. 1993. The electrochemical dissolution of gold in bromine medium. *Minerals Engineering.* 6: 1067–1079.

von Michaelis, H. 1987. The prospects for alternative leach reagents (for gold). *Engineering and Mining Journal.* 188: 42–47.

Narita, H., Tanaka, M., Morisaku, K., Abe, T. 2006. Extraction of gold(III) in hydro-chloric acid solution using monoamide compounds. *Hydrometallurgy.* 81(3–4): 153–158.

Nicol, W. M. 1976. Production of crystalline sugar U.S. Patent No. 3, 972, 725. U.S. Patent and Trademark Office, Washington, DC.

Nicol, M.J. 1980. The anodic b›aviour of gold. *Gold Bulletin.* 13(2): 46–55.

Nicol, M.J., Paul, R.L., Fleming, C.A. 1987. The chemistry of the extraction of gold. *Mint–.* http://www.saimm.co.za/Conferences/ExtractiveMetallurgyOfGold/0831-Chapter15.pdf.

Pesic, B., Sergent, R.H. 1992. Dissolution of gold with bromine from refractory ores pre-oxidized by pressure oxidation. In: Hager, J. (Ed.), *Proceedings, EDP '92 Congress*. The Minerals, Metals and Materials Society, Warrendale, PA, pp. 99–114.

Putnam, G.L. 1944. Chlorine as a solvent in gold hydrometallurgy. *Engineering and Mining Journal*. 145(3): 70–75.

Puvvada, G.V.K., Murthy, D.S.R. 2000. Selective precious metals leaching from a chalcopyrite concentrate using chloride/hypochlorite media. *Hydrometallurgy*. 58: 185–191.

Radulescu, O., Gorban, A.N., Zinovyev, A., Lilienbaum, A. 2008. Robust simplifications of multiscale biochemical networks. *BMC Systems Biology*. 2(1), 86.

Sergent, J., Ohta, S., Macdonald, B. 1992. Functional neuroanatomy of face and object processing: A positron emission tomography study. *Brain*. 115(1): 15–36.

Severs, K. 1999. Technological advances in treating copper concentrates—The Intec copper process. Paper presented at the 128th SME Meeting, Denver, CO, March 1–3, 1999.

Snoeyink, P.L., Jenkins, D. 1979. *Water Chemistry*. John Wiley & Sons, New York.

Sparrow, G.J., Woodcock, J.T. 1995. Cyanide and other lixiviant leaching systems for gold with some practical applications. *Mineral Proceccsing and Extractive Metallurgy Reviews*. 14: 193–247.

Tran, T., Davis, A., Song, J. 1992a. Extraction of gold in halide media. In: Mishra, V.N., Halbe, D., Spottiswood, D.J. (Eds), *Proceedings of International Conference on Extractive Metallurgy of Gold and Base Metals*, The AuSIMM, Kalgoorlie, WA, pp. 323–327.

Tran, P., Zhang, X.K., Salbert, G., Hermann, T., L›mann, J.M., Pfahl, M. 1992b. COUP orphan receptors are negative regulators of retinoic acid response pathways. *Molecular and Cellular Biology*. 12(10): 4666–4676.

Tran, T., Lee, K., Fernando, K. 2001. Halide as an alternative lixiviant for gold processing—An update. In: Young, C.A., Twidwell, L.G., Anderson, C.G. (Eds), *Cyanide: Social, Industrial and Economic Aspects*. The Minerals, Metals and Materials Society, Warrendale, PA, pp. 501–508.

Trindade, R.B.E., Rocha, P.C.P., Barbosa, J.P. 1994. Dissolution of gold in oxidized bromide solutions. In: *Hydrometallurgy '94*. Chapman & Hall, London, pp. 527–540.

Viñals, J., Nunez, C., Herreros, O. 1995. Kinetics of the aqueous chlorination of gold in suspended particles. *Hydrometallurgy*. 38(2): 125–147.

Yen, W.T., Pindred, R.A., Lam, M.P. 1990. Hypochlorite leaching of gold ore. In: Hiskey, J.B., Warren, G.W. (Eds), *Hydrometallurgy Fundamentals, Technology and Innovations*. The Society for Mining, Metallurgy and Exploration Inc., Littleton, CO, pp. 415–436.

7

Microbial Cyanidation of Gold

Sadia Ilyas* and Nimra Ilyas†

7.1 Microbial Role in Gold Metallurgy

The previous chapters described pretreatment followed by gold extraction processes with various chemical reagents basically used for liberating or dissolving the precious metal from its native state via complexation with the employed chemical. Cyanide, thiourea, thiosulfate, and halides are the prominent chemicals used for dissolving gold with pretreatment steps (used for refractory ores) like oxidation, roasting, autoclaving, etc. The environmental footprints of almost all of these chemicals are a matter of consideration on which basis the precious gold has been named as the "dirty gold" – although thiourea, thiosulfate, and iodide leaching are considered less toxic (Lee and Srivastava, 2016). With the stringent environmental rules and increasing public awareness regarding health and environment issues, it is imperative to search for an alternative route that could change the narrative of "dirty gold" to "greener gold." As compared to the conventional chemical route, application of microbial activity in the field of metallurgy has been considered a greener, low-cost technology (Ilyas and Lee, 2014; Ilyas et al., 2017a,b). Significant interest has been shown in the past few years for developing and commissioning the microbial process to assist the commercial recovery of gold, in particular from the refractory ores. Figure 7.1 depicts a brief classification of organisms. The potential role of microbial activity in gold metallurgy can be divided into four steps (Dew et al., 1999; Edwards et al., 2000):

i. pretreatment via bio-oxidation,
ii. leaching of gold in biogenic lixiviants,
iii. bio-recovery following bio-sorption, and
iv. microbial treatment of the effluents generated in the process.

---

* Mineral and Material Chemistry Lab, Department of Chemistry, University of Agriculture Faisalabad, Pakistan.
† Institute of Microbiology, University of Agriculture Faisalabad, Pakistan.

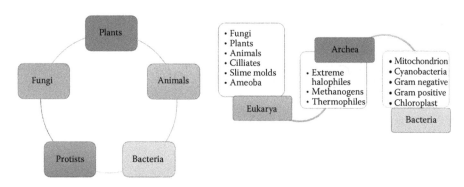

FIGURE 7.1
Classification of microorganisms.

All the aspects of microbial activities are discussed in this chapter after a brief discussion of the microbes-to-metal interactions.

7.2 Microbes-to-Metal Interactions

The microorganisms encounter metal-constituent bodies in their environment, either to their benefit or detriment, and depend on a system that can be single- or multi-metal and the types of microbes that can be prokaryotic or eukaryotic. Both types of microbes can act to (i) bind the metal species which influences the cell surface; (ii) carry them into the cell to produce various intercellular functions; or (iii) form various metabolic products which can be used for chelating the metals (Figure 7.2). Furthermore, the oxidation of refractory ores and removal of disrupting ore-constituents is an act following either the direct or indirect mechanism, or a combination of both.

The mechanism of microbial activities has further been refined by Crundwell (2003) as:

i. the attached microbes on the mineral surface oxidize the mineral bodies (usually sulphides) with a release of H_2SO_4;

ii. the attached microbes oxidize the metal ions (e.g. Fe^{2+} to Fe^{3+}) to mediate the metal liberation from mineral (usually sulphide) phases; and

iii. the transfer of oxidized ions (Fe^{3+}) to the bulk solution mediating the reactions onto mineral surface.

An example of a direct mechanism is enzymes mediated (oxidases/reductases), whereas an indirect mechanism is associated with metabolites (organic/inorganic acids) of the microorganisms (Dew et al., 1999); however,

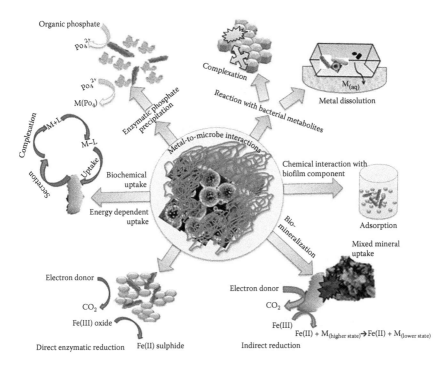

FIGURE 7.2
Various modes of microbial interactions with metals (Ilyas et al., 2017a).

no strong evidence exists for a direct breaking of sulphides by microbial activities (Sand et al., 1995; Tributsch, 2001; Watling, 2006). The most proven mechanism is understood by the thiosulfate or polysulphate pathways, as shown in Figure 7.3 (Rawlings, 2005; Erüst et al., 2013; Ilyas et al., 2014, 2017b).

After the microbial pretreatment (usually termed as bio-oxidation) step, leaching of gold with suitable lixiviant can be performed, either with chemical or biogenic reagents. The chemical leaching has previously been described; hence, in this chapter only the subsequent microbial leaching of gold is considered.

7.3 Bio-Oxidation of Gold-Bearing Refractory Ores

With a fast depletion of higher grade free-milling ores, the processing of gold-bearing refractory ores is becoming increasingly attractive. Nevertheless, in a refractory ore gold remains locked in the sulphide minerals and siliceous gangue or sometimes with carbonaceous materials, causing inefficient leaching of gold and therefore requires treating with specific care, often with a

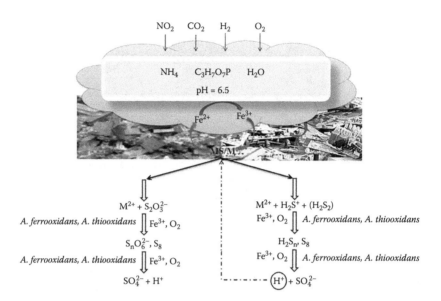

FIGURE 7.3
Thiosulphate and polysulphide pathway of biooxidation (Ilyas et al., 2017a).

pretreatment step (Ilyas et al., 2010, 2014; Ilyas and Lee, 2014). As the gold mining and metallurgical industries are se–ing a flexible, robust, cost-effective, and environmentally friendly technique, bio-oxidation has been widely applied from laboratory to commercial scales, including the Harbour Lights Mine (Australia), Fairview Mine (South Africa), and Sao Bento (Brazil). The Newmont Gold Company (USA) is a prominent example using bio-oxidation in a heap of a million tons sulphide minerals in Nevada. The favourable effects of using bacteria, archaea, or fungi in metal extraction from mineral bodies are mainly based on the principles of:

 i. acidolysis – the mobilization of metals with organic/inorganic acid/s;

 ii. Redoxolysis – the mobilization of metals using redox mechanisms; and

 iii. Complexolysis – the mobilization of metals using the biogenic complexing agents.

7.3.1 Microorganisms in Bio-Oxidation

The sulphide minerals oxidized by the microorganism are able to form soluble metal sulphates and produce sulfuric acid. The mesophilic *Thiobacillus ferrooxidans* can oxidize ferrous iron and reduce the sulphur compound of the pyrite, chalcopyrite, or arsenopyrite mineral; whereas, *Leptospirillum ferrooxidans* only oxidizes the ferrous iron. These microorganisms develop prominently in an ambient environment, usually in mine sites of sulphide ores. Table 7.1 provides the list of bacterial, archaeal, and fungus species that are most commonly used in bio-oxidation. They utilize the iron as their energy source (oxidizing Fe^{2+} to Fe^{3+} ions and consuming elemental sulphur to convert it in H_2SO_4 provides the proton and oxidant to attack on mineral surfaces), contributing to indirect leaching of sulphide minerals. However, most of the microorganisms meet their carbon requirements by utilizing the gas phase of CO_2, leaving a substantial amount of carbonaceous solid that may affect the subsequent leaching due to preg-robbing of gold (Yen et al., 2008; Ofori-Sarpong et al., 2011; Ilyas and Lee, 2015). Reaction is initiated at the membrane bonded cytochromes and moves through the up-hill and down-hill pathway (Figure 7.4). But a recent study using the fungal strain *Phanerochaete chrysosporium* showed that it could reduce the preg-robbing effect of anthracite-grade carbonaceous matter (Ofori-Sarpong et al., 2010). The results revealed its ability to reduce 15%–35% of sulphide compounds, which subsequently increased the overall efficiency of gold leaching (Ofori-Sarpong et al., 2011). The following factors make the overall bio-oxidation process more efficient:

Microbial-based factors: The distributed microbial diversity, population density, adopted (percolate/suspended) leaching methods, and tolerance level to metal concentration affect the overall bio-oxidation efficiency (Das et al., 1999).

Metabolite-based factors: The formation of extracellular polysaccharides plays a very important role in the contaminant removal from refractory mineral ores. Loss of these exopolymers along with the release of carbonaceous matters adversely affects the overall efficiency (Rohwerder et al., 2003).

Reaction-based factors: The physic-bio-chemical factors (like temperature, pH, nutrients, particle size, metal species, absence/presence of oxygen, pulp density, heavy metals, and mineral composition) affect the overall bio-oxidation and gold leaching efficiency (Bosecker, 1997; Mousavi et al., 2005). The mineralogical characteristics of gold bearing ores significantly affect the rate of bio-oxidation and the extent to which the ore can be oxidized. For example, arsenopyrite can preferentially be oxidized in a mixed pyrite-arsenopyrite system; hence, gold primarily associated with the arsenopyrite can be liberated by oxidizing only a fraction of the total sulphide minerals. Notably, different rates of bio-oxidation may occur even among the arsenopyrites, depending on their crystal structures. Using the heterotrophic bacterial consortium (of species *Pseudomonas, Arthrobacter,* and *Achromobacter*) a partial blocking of preg-robbing effect has been reported (Kulpa and Brierley, 1993). Such a blocking may be due to the formation of a biofilm and/or the adsorption of

TABLE 7.1

Prominently Used Arceal, Bacterial and Fungus Species for Biooxidation

Organism	Nutrition Type	Domain	Energy Source/ Biooxidation	pH	Temperature
Acidianus ambivalens	Facultative heterotrophic	Archea	Sulfur/heavy metals	Acidophilic	Thermophiles
Metallosphaera prunae	Chemolithoautotrophic	Archea	Iron, sulphur/heavy metals	Acidophilic	Moderate thermophiles
Acidianus brierleyi	Facultative heterotrophic	Archea	Sulfur/heavy metals	Acidophilic	Moderate thermophiles
Sulfobacillus thermosulfidooxidan		Archea	Iron, sulphur/heavy metals		
Ferroplasma acidiphilum	Chemolithoautotrophic	Archea	Iron/heavy metals	Acidophilic	Moderate thermophiles
Acidianus infernus	Facultative heterotrophic	Archea	Iron, sulphur/heavy metals	Acidophilic	Moderate thermophiles
Metallosphaera sedula	Chemolithoautotrophic	Archea	Iron, sulphur/heavy metals	Acidophilic	Extreme thermophilic
Picrophilus oshimae	Chemolithoautotrophic	Archea	Iron, sulphur/heavy metals	Acidophilic	Extreme thermophilic
Picrophilus torridus	Chemolithoautotrophic	Archea	Iron, sulphur/heavy metals	Acidophilic	Extreme thermophilic
Sulfolobus acidocaldarius	Chemolithoautotrophic	Archea	Iron, sulphur/heavy metals	Acidophilic	Extreme thermophilic
Sulfolobus ambivalens	Chemolithoautotrophic	Archea	Iron, sulphur/heavy metals	Acidophilic	Extreme thermophilic
Sulfolobus brierleyi	Chemolithoautotrophic	Archea	Iron, sulphur/heavy metals	Acidophilic	Extreme thermophilic
Sulfolobus hakonensis	Chemolithoautotrophic	Archea	Iron, sulphur/heavy metals	Acidophilic	Extreme thermophilic
Sulfolobus thermosulfidooxidans	Chemolithoautotrophic	Archea	Iron, sulphur/heavy metals	Acidophilic	Extreme thermophilic
Sulfurococcus mirabilis	Chemolithoautotrophic	Archea	Iron, sulphur/heavy metals	Acidophilic	Extreme thermophilic
Sulfolobus yellowstonii	Mixotrophic	Archea	Iron, sulphur/heavy metals	Acidophilic	Extreme thermophilic
Thermoplasma acidophilum	Mixotrophic	Archea	Iron, sulphur/heavy metals	Acidophilic	Extreme thermophilic

(Continued)

TABLE 7.1 (Continued)

Prominently Used Arceal, Bacterial and Fungus Species for Biooxidation

Organism	Nutrition Type	Domain	Energy Source/ Biooxidation	pH	Temperature
Acidimicrobium ferrooxidans	Heterotrophic	Bacteria	Glucose/heavy metals	Acidophilic	Mesophile
Acidocella sp.	Heterotrophic	Bacteria	Glucose/heavy metals	Acidophilic	Mesophile
Thiobacillus albertis	Chemolithcautotrophic	Bacteria	Sulfur/heavy metals	Acidophilic	Mesophile
Thiobacillus acidophilus	Mixotrophic	Bacteria	Sulfur/heavy metals	Acidophilic	Mesophile
Acidomonas methanolica	Heterotroph	Bacteria	Glucose/heavy metals	Acidophilic	Mesophile
Thiobacillus thiooxidans	Chemolithoautotrophic	Bacteria	Sulfur/heavy metals	Acidophilic	Mesophile
Coriolus versicolor		Fungi	Glucose/heavy metals	Mild acidophilic	Mesophile
Aspergillus avamori	Heterotroph	Fungi	Glucose/heavy metals	Mild acidophilic	Mesophile
Aspergillus fumigatus	Heterotroph	Fungi	Glucose/heavy metals	Mild acidophilic	Mesophile
Aspergillus niger	Heterotroph	Fungi	Glucose/heavy metals	Mild acidophilic	Mesophile
Aspergillus ochraceus	Heterotroph	Fungi	Glucose/heavy metals	Mild acidophilic	Mesophile
Aspergillus sp.	Heterotroph	Fungi	Glucose/heavy metals	Mild acidophilic	Mesophile
Fusarium sp.	Heterotroph	Fungi	Glucose/heavy metals	Mild acidophilic	Mesophile
Cladosporium sp.	Heterotroph	Fungi	Glucose/heavy metals	Mild acidophilic	Mesophile

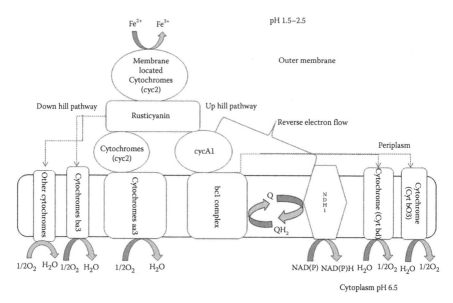

FIGURE 7.4
Iron oxidation pathway in *Acidithiobacillus ferooxidans*.

excreted extracellular organic products onto mineral surfaces, which could passively block the contact between the gold-cyanide complex and carbonaceous matter.

The scale of bio-oxidation along with the use of bioengineering aspects also affects the oxidation rate. Stirred tank reactors provide an efficient mass transfer rate of the supplied gases (O_2 and CO_2) to the bio-oxidation system. When the fine grinding of low grade ore in large amounts is uneconomical, the bio-oxidation in heaps is being practiced. Crushed ore sized ~20 mm is generally used to provide the easy gravity-flow of lixiviant and supplied air (that can enhance the rate of bio-oxidation) to the heap.

Notably, at commercial scale operations the best performance of bio-oxidation has been noticed in continuous stirred tank reactors with high volumetric productivity. In the last two decades, several commercial mineral processing units have been established using either the BIOX (Bio-oxidation) or BacTech (Bacterial Oxidation Technology) processes. A BIOX plant typically operates up to a maximum 18%–20% pulp density with a total retention time of around four days. The alternate BacTech (Australia) process uses similar aerated stirred tank reactors; however, it uses a significant variance of operating temperature, elevated to ~50°C. Hence, it uses moderately thermophilic bacteria with the external addition of growth nutrients (nitrogen, phosphorous, and potassium) that can vary with changing concentration of gold-bearing concentrates.

7.4 Bio-Cyanidation and Cyanogenesis

Harmful environmental footprints due to a large amount of cyanide used in gold leaching from ores have created the need to search for an alternative without compromising recovery efficiency. Cyanide has a proven record of higher leaching rate and recovery efficiency with a low-cost carbon adsorption technique and has yet to be replaced by another reagent on a large, commercial scale. Therefore, bio-cyanidation is gaining much attention in recent times by the use of a variety of cyanogenic microorganisms. The *in* situ biogenesis of cyanide in or nearby gold deposits by microorganisms can be applied to leach out gold with minimal environmental risks. Other than cyanide, however, the amino acids—serine, asparagine, histidine, aspartic acid, glycine, and alanine – are also capable of forming the soluble gold complexes in an alkaline medium (Olson, 1994), but only a few milligrams of gold can be yielded.

The biogenic cyanide is produced only for a short duration in an early stationary phase in the presence of glycine (Castric, 1981; Merchant, 1998; Kita et al., 2006) as a secondary metabolite due to its independent production from the growth phase (Castric, 1975). Glycine (NH_2CH_2COOH) acts as a precursor that is formed by an oxidative decarboxylation catalysed by the enzyme HCN-synthase and is associated mainly with the membrane of the microorganisms belonging to the cyanogenic groups. Glycine gets oxidized to imino acetic acid (H–C(NH)–COOH) followed by the split of the C–C bond with a concomitant second dydrogenase reaction that produces HCN and CO_2 (Laville et al., 1998), as in Figure 7.5. Thus, the produced cyanide is volatile in nature and can be reduced in the presence of salts and other cyanicidic compounds in the growth medium (Faramarzi and Brandl, 2006).

The cyanogenic bacterial species like *Chromobacterium violaceum*, *Pseudomonas fluorescens*, *Pseudomonas aureofaciens*, *Pseudomonas aeruginosa*, *Pseudomonas plecoglossicida*, *Pseudomonas putida*, *Pseudomonas syringae*, *Bacillus megaterium*, archaea species viz. *Ferroplasma acidipholum*, and *Ferroplasma acidarmanus*, and some fungal species such as *Marasmius oreades*, *Clitocybe sp.*,

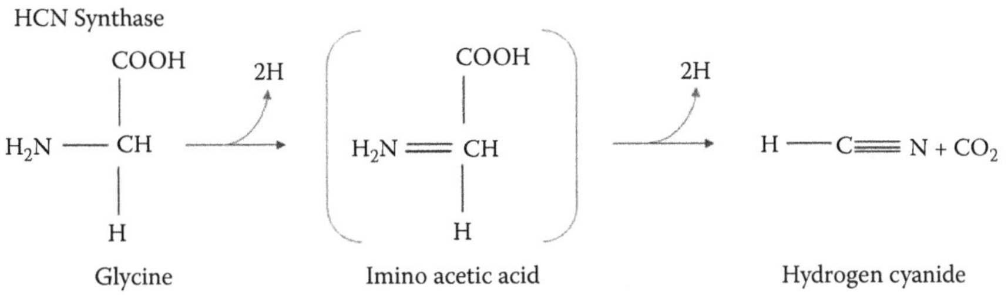

FIGURE 7.5
HCN synthase of glycine with intermediate formation of imino acetic acid.

and *Polysporus* sp. have been reported so far (Knowles, 1976; Askeland and Morrison, 1983; Paterson, 1990; Flaishman, 1996; Golyshina et al., 2000; Faramarzi et al., 2004; Faramarzi and Brandl, 2006; Brandl et al., 2008; Hol et al., 2011). However, the application of fungi and archaea is limited, and commercially available microbial isolates (*Chromobacterium violaceum* and *Pseudomonas fluorescens*) are the most commonly studied.

7.4.1 Bio-Cyanidation by *Chromobacterium violaceum*

Chromobacterium violaceum is a Gram-negative, violet coloured, rod-shaped, facultative anaerobic coccobacillus. This is a soil and water organism, which is common in tropical and subtropical countries and may cause serious pyogenic or septicemic infection in mammals. *Chromobacterium violaceum* is a mesophilic that can grow at temperature between 10°C and 40°C and around a neutral pH of 7–8, producing violacein in the presence of trypto-phan. It ferments glucose, N-acetylglucosamine, gluconate, and tr›alose and is found to be positive for oxidase and catalase reactions. The maxi-mum cyanide production by *Chromobacterium violaceum* occurred in the onset of the stationary phase (Castric, 1975; Laville et al., 1998; Lawson et al., 1999; Kita et al., 2006). Bio-leaching of gold using *Chromobacterium violaceum* is highly dependent on variables such as ore type and gold content in the ore bodies.

7.4.2 Bio-Cyanidation by *Pseudomonas fluorescens*

Cyanogenesis by *Pseudomonas fluorescens* has been studied to a greater extent than by *Chromobacterium violaceum* (Campbell et al., 2001). It is the genus of aerobic *Pseudomonas* family *Pseudomonadaceae* (Kita et al., 2006). Like *Chromobacterium violaceum*, it is also rod-shaped, becoming shorter and thinner in old cultures. It is motile with polar multitrichous flagellation; however, it is nonmotile in some cases. It produces diffusible fluorescent pigment, particularly in iron-deficient media, and is able to use >80 differ-ent carbon substrates during the growth period, usually at 25°C–30°C. The optimal pH for cyanogenesis is 8.3 in Tris/HCl buffer; however, the optimal pH in other buffers is in between 7.3 and 7.8.

7.4.3 The Metabolic Pathway

Glycine mainly acts as an immediate metabolic precursor of cyanide in the proteobacteria, and a stoichiometric formation of HCN and CO_2 occurs via the oxidative decarboxylation of it (Michaels et al., 1965; Wissing, 1974; Castric, 1977). Using the substrate of radiolabelled [1–14C] glycine or [2–14C] glycine, cyanide can be derived from the methylene carbon of glycine and CO_2 from the carboxyl group of glycine in the cyanogenic bacteria *Chromobacterium violaceum*, *Pseudomonas fluorescens*, and *Pseudomonas aeruginosa* (Bunch and

Knowles, 1982; Askeland and Morrison, 1983). The C–N bond is retained during the reaction (Brysk et al., 1969) without any possible transamination or deamination reactions. In *Pseudomonas aeruginosa*, threonine can be metabolized to glycine to serve as a precursor for biocyanidation, but less effectively (Castric, 1977). In vitro, the enzyme complex converting glycine to HCN and CO_2, HCN synthase appears to be membrane-associated, both in a *Chromobacterium violaceum* and *Pseudomonas* sp. that can be solubilized by detergents (Wissing, 1975; Wissing and Andersen, 1981; Bunch and Knowles, 1982). Notably, the HCN synthase is very sensitive to oxygen and can be altered easily; however, glycine somewhat protects the *Pseudomonas* and *Chromobacterium* enzymes from the toxicity exhibited by oxygen (Castric, 1981; Wissing and Andersen, 1981; Bunch and Knowles, 1982). Hence, the HCN synthase has been purified only partially from a *Pseudomonas* sp. In crude extracts, HCN synthase of *Pseudomonas* sp., and *Chromobacterium violaceum* can use artificial electron acceptors to oxidize glycine (Wissing, 1975; Castric, 1981; Bunch and Knowles, 1982). Whereas in vivo the natural electron acceptor is oxygen, it is not strictly required for cyanogenesis by the *Chromobacterium violaceum* in which fumarate can be the terminal electron acceptor.

7.5 Bioleaching of Gold in Microbial Produced Cyanide Solution

The biogenic cyanide, produced as previously defined, can be utilized for leaching the gold-bearing ores (after pre-bio-oxidation while processing the refractory ores) into it, as shown in Figure 7.6. Although the leaching of gold in cyanide via complexation is similar to the chemical cyanidation leaching, the overall bio-cyanidation leaching of gold can be divided into three types of processes: (i) one-step bioleaching, (ii) two-step bioleaching, and (iii) spent medium bioleaching; mainly based on whether the ore is added directly or indirectly into the media, or cells are separated from the culture after reaching maximum cell density, and cyanide production and ore is added into the cell free media (spent media). A prior decrease in copper contents in ore bodies is preferentially suggested to achieve an improved bio-leaching of gold (Rohwerder et al., 2003). An external addition of $FeSO_4$ and $MgSO_4$ to the medium and presence of Na_2HPO_4 and $Pb(NO_3)_2$ is found to enhance the rate of cyanogenesis (Bosecker, 1997; Brandl et al., 2001; Hagelüken, 2006; Hoque and Philip, 2011; Ilyas and Lee, 2014).

In contrast to a two-step bioleaching where microbes consume the oxygen, the spent medium bioleaching utilizes oxygen to be mixed with gold in the absence of bacteria. As no consumption of cyanide is done by the bacteria in

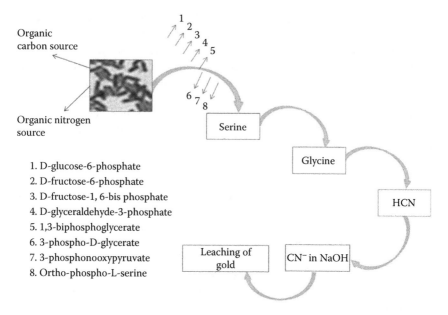

Organic carbon source

Organic nitrogen source

1. D-glucose-6-phosphate
2. D-fructose-6-phosphate
3. D-fructose-1, 6-bis phosphate
4. D-glyceraldehyde-3-phosphate
5. 1,3-biphosphoglycerate
6. 3-phospho-D-glycerate
7. 3-phosphonooxypyruvate
8. Ortho-phospho-L-serine

Serine

Glycine

HCN

Leaching of gold

CN^- in NaOH

FIGURE 7.6
An entire schematic for bioleaching of gold with microbial produced cyanide solution (Ilyas et al., 2017b).

spent media leaching, the full strength of biogenic cyanide can be utilized in complexing with gold as compared to the two-step bioleaching where growth/cyanide production is coupled with leaching process (Bulmer, 2000). A significant amount of metals are found to be bioaccumulated and biosorped in two-step bioleaching. Parallel to the metal-cyanide complexation, bacteria continue to bioaccumulate gold (along with other metals in the system), which reduces the leaching yield in a two-step leaching system. It also reveals that only the metabolites (which is cyanide) produced by *Chromobacterium violaceum* are involved in the bioleaching process. The advantage of spent medium bioleaching is the elimination of pulp density limitation, because the leaching is performed separately from the microbial growth vessel or tank, which prevents the toxicity caused by higher metal concentration. This makes the operation possible at higher pH levels than are suitable for cyanide stability.

7.6 Bio-Recovery of Gold from Leach Liquors

The recovery of gold from the leach liquors obtained by bio-cyanidation can be carried out the same as from chemical cyanidation leach liquor (see Chapter 3). In this chapter, only the possibility of gold recovery by biological/microbial activities (usually following the adsorption phenomena) is discussed.

The passive sorption of gold onto the surfaces of the biomass and/ or the complexation of gold ions with the surface functional groups of the particular biomass can be termed as the biosorption in gold metallurgy. Bacteria, algae, and fungi have shown proficient and economical recovery of gold; along with brown alga, *Fucus vesiculosus*, yielding metallic gold nano-particles by (Torres et al., 2005). The brown marine alga, *Sargassumnatans*, and green algae, *Chlorella vulgaris*, have also shown high discrimination for gold adsorption (Greene et al., 1986; Kuyucak and Volesky, 1988). Fungi such as *Aspergillus niger, Mucor rouxii* and *Rhizopus arrihus* can selectively adsorb gold (Townsley and Ross, 1986; Mullen et al., 1989), while the fungus strains *Cladosporioum cladosporoides* 1 and 2 showed better results with dilute solutions (Pethkar and Paknikar, 1998). Bracket fungi (*Fomitopsis carnea*) immobilized in polyvinyl alcohol have shown an adsorption capacity of 94 mg gold/g biomass (Khoo and Ting, 2001). Bacteria strains *Streptomyces phaeochromogenes* HUT6013 showed the highest bio-sorption of gold (282 µmol/g dry weight cells). Numerous species of yeast (*Candida krusei, Candida robusta, Candida utilis, Cryptococcus albidus, Cryptococcus laurentii, Debaromyces hansenii, Endomycopsisfibigera, Hansenula anomala, Hansenula saturnas, Kluyveromyces Pichia farinose, Saccharomyces cerevisiae, Sporobolomyces salmonicolor* and *Torulopsisaeria*) are able to efficiently adsorb gold ions from dilute solutions (Tsuruta, 2004; Lin et al., 2005) due to their favourable oxygenous functional groups onto the cell wall. High affinity for biosorption of gold by Gram negative bacterial strains (*Acinetobacter calcoaceticus, Erwinia herbicola, Pseudomonas aeruginosa* and *Pseudomonas maltophilia*) in the presence of hydrogen tetrachloroaurate has been reported (Tsuruta, 2004). Animal products, like a hen egg shell membrane from electroplating wastewater (Ishikawa et al., 2002), alfalfa biomass (Gamez et al., 2003), chemically modified hop biomass (López et al., 2005), persimmon peel gel (Parajuli et al., 2007), rice husk carbon and barley straw carbon (Chand et al., 2009), cross-linked chitosans (Ahmad et al., 2003; Arrascue et al., 2003; Fujiwara et al., 2007), activated carbon from hard shell Iranian apricot stones (Soleimani and Kaghazchi, 2008), and Diethylaminoethyl-cellulose (DEAE-cellulose) (Tasdelen et al., 2009) are some of the prominently employed names of bio-sorbents used to recover gold from divergent solutions. Table 7.2 summarizes some of the bio-sorbents with their critical conditions employed to adsorb gold from divergent solutions.

Notably, a successful desorption process for efficient recovery of gold would require a proper selection of elutant and is strongly dependent on the type of bio-sorbent used and the mechanism it follows. It would also need to be nondestructive to the biomass, cost-effective, and eco-friendly. For example, desorption from egg shell membrane is prominent with 0.1 mol/L NaOH and NaCN solution (Ishikawa et al., 2002), while the elution of gold from brown marine alga (*Saragasum natans*) yields better with HCl solutions.

TABLE 7.2

Summary of Various R&D Works Done for Recovering the Gold Using Biosorption Techniques

Microbe Used (Fungi + Yeast)	Metal	Optimum Process Conditions	Recovery	Remarks	References
Cladosporium cladosporioides	Synthetic solution containing Au^{2+}	pH 4, contact time 30 min, temp 30°C, biosorbent 0.05 g/25 mL	Au 96.6 mg/g Ag 44.5 mg/g	Colloidal form of gold	Pethkar and Paknikar (1998)
Cladosporium cladosporioides	Synthetic solution containing Ag^+	pH 4, contact time 30 min, temp 30°C, biosorbent 0.05 g/25 mL	Au 105 mg/g Ag 15.2 mg/g	Colloidal form of gold	Pethkar and Paknikar (1998)
Rhizopus arrhizus	8.5–1000 mg Au in chloride solution with UO_2^{2+}, Pb^{2+}, Zn^{2+}, Ag^+	pH 2.5, contact time 30 min, temp 30°C, biomass 0.5 mg/mL	92%–95% Au,	Colloidal gold	Kuyucak and Volesky (1988)
PVA-immobilized biomass (*Fomitopsis carnea*)	Gold solution 10–100 mg dm^{-3} Au	pH 1–13, contact time 33.5 h, temp 25°C, biomass 5 g/100 mL of gold solution	80% Au	Colloidal gold	Khoo and Ting (2001)
Algae					
Chlorella vulgaris	A solution: 10^{-4} M of each of Cr^{3+}, Ag^{2+}, Cu^{2+}, Zn^{2+}, Au^{3+} Hg^{2+}	Contact time 12 h, pH 2, temp 30°C, biomass 5 mg/mL	90%	Gold as $AuCl_4$ and Cr as $Cr(H_2O)_6^{3+}$, other metals as oxides	Darnall et al. (1986)
Sargassum natans	Gold solution: 8.5–1000 mg gold chloride with other metal ions UO_2^{2+}, Pb^{2+}, Zn^{2+}, Ag^+	pH 2.5, contact time 2 h, temp 30°C	420 mg/g gold	Cationic species of gold	Kuyucak and Volesky (1988)
Sargassum fluitans	Synthetic solution: 2.2 mM gold cyanide	pH 2, contact time 4 h, temp 30°C, biomass 20 mg/150 mL metal solution	97%	Anionic gold species	Niu and Volesky (1999)

(Continued)

TABLE 7.2 (Continued)

Summary of Various R&D Works Done for Recovering the Gold Using Biosorption Techniques

Microbe Used (Fungi + Yeast)	Metal	Optimum Process Conditions	Recovery	Remarks	References
Alginate cross-linked with $CaCl_2$	Synthetic solution of $HAuCl^4$: 25–500 ppm gold	pH 2, contact time 4 h, temp 25°C, biomass 0.075 g/75 mL metal solution	98%	Colloidal gold	Torres et al. (2005)
Dealginated seaweed waste	Synthetic solution of gold	pH 3, temp 25°C, biomass 0.5 g/mL metal solution	92%	Colloidal gold	Romero-González et al. (2003)
Protein					
Hen egg shell membrane	Synthetic solution: 1000 mg/L $KAu(CN)_2$, $HAuCl_4H_2O$	pH 3, contact time 2 h, temp 25°C, biosorbent 0.25–0.035 g/50 mL gold solution	147 mg Au^{1+}/g 618 mg Au^{3+}/g	$Au(CN)_2$, $Au(CN)_4$	Ishikawa et al. (2002)
Lysozyme from hen egg white	Copper refining solution: 82 g/L Au	pH 4, Contact time 1 h, temp 25°C, biosorbent 2 mg/10 mL	165 g/Kg Au	Gold ions	Maruyama et al. (2007)
Alfalfa	Aqueous solution: 0.3 mM of the each of metal ions: Au^{3+}, Cd^{2+}, Cu^{2+}, Cr^{3+}, Pb^{2+}, Ni^{2+}, Zn^{2+}	pH 5, contact time 60 min, temp 30°C, biomass 5 mg/mL	58% Au	Metals in oxidized form	Gamez et al. (2003)
Acid-washed *Ucides cordatus* (waste crab shells)	Synthetic solution: 2.2 mM gold cyanide	PH 3.4, contact time 24 h, temp 25°C, biosorbent 40 mg/20 mL gold solution	92% Au	$Au(CN)_2$, $Au(CN)_4$	Niu and Volesky (2003)

(Continued)

TABLE 7.2 (Continued)

Summary of Various R&D Works Done for Recovering the Gold Using Biosorption Techniques

Microbe Used (Fungi+Yeast)	Metal	Optimum Process Conditions	Recovery	Remarks	References
Glutaraldehyde crosslinked derivatives of chitosan (GCC, RADC)	Synthetic gold solution: HAuCl₄ in 1M HCl)	Biosorption/reduction (dual), pH 2–3, contact time 5 days, Temp ~30°C, biosorbent 8 mg/150 mL gold	600 mg/g Au	Gold as hydroxide and chloride forms	Arrascue et al. (2003)
Sulfur derivative of chitosan	Synthetic gold solution: HAuCl₄, in 1M HCl	pH 3, contact time 5 days, temp 30°C, biosorbent 8 mg/150 mL gold	400 mg/g Au	Gold hydroxide and chloride species	Arrascue et al. (2003)
Bacteria					
Streptomyces erythraeus	Synthetic solution: 1000 mg/L KAu(CN)₂, HAuCl₄H₂O	pH 3, contact time 2 h, temp 25°C, biosorbent 0.25–0.035 g/50 mL gold solution	99% Au	Au(CN)₂, Au(CN)₄	Ishikawa et al. (2002)
Spirulina platensis	Synthetic solution: 1000 mg/L KAu(CN)₂, HAuCl₄H₂O	pH 3, contact time 2 h, temp 25°C, biosorbent 0.25–0.035 g/50 mL gold solution	98% Au	Au(CN)₂, Au(CN)₄	Ishikawa et al. (2002)
Bacillus subtilis	Synthetic solution: 2.2 mM gold cyanide	pH 2, contact time 4 h, temp 30°C, Biosorbent 40 mg/20 mL gold solution	97% Au	Au(CN)₂, Au(CN)₄	Niu and Volesky (1999)

7.7 The Microbial-Mediated Destruction of Cyanide

The adversity caused by the free cyanide abundantly discharged in the environment is discussed in detail in Chapter 3 (related to the chemical cyanidation of gold-bearing ores). The use of bio-cyanidation is a prominent solution of the problems created by chemical cyanidation:

i. Cyanogenic bacteria can produce *in* situ cyanide as a secondary metabolite in an early stationary phase that can be utilized in gold leaching from the ore bodies; and

ii. The breakdown and/or transformation of residual cyanide (remained in effluents to be discharged) to simpler and less toxic products in the latter stage of their life span, by the microbial activity of the species used in cyanogenesis.

This latter stage process exploits the diversity in metabolic modes and adaptability of microorganisms in cyanide degradation. The bio-degradation of cyanide is based on the fact that microbes use cyanide as an energy source to get nitrogen and carbon for the synthesis of amino acids and oxidize the cyanide to convert into products like carbonate and ammonia (Adams et al., 2001).

The following are identified advantages for employing the bio-degradation of cyanide:

i. All types of cyanide (including ionic cyanide, Weak Acid Dissociable (WAD) cyanide, ferrocyanides, thiocyanides, thiocyanates, and metal-cyanide complexes) can be treated by this method.

ii. Different microbes are available with optimum degradation rates from 4°C to 30°C.

iii. It can be applied in various stages of operation and sources of contamination.

iv. Treatment of waste containing high CN concentrations >350 mg/L at wider pH range of 7.5–11.5 is possible.

v. It is a low-cost operation that can be adapted to existing mine infrastructure with nutrients costing ~0.05 USD/1,000 gallons.

The bio-degradation path of cyanide involves a combination of hydrolytic, oxidation, reduction, and substitution processes.

7.7.1 Oxidative Pathway

Many of the enzymes able to catalyse the transfer of electrons from one molecule to another (oxidoreductases) follow the oxidative pathway to yield the by-products ammonia and carbon dioxide. First, the monoxygenase catalyses the cyanides to cyanates and later hydrolyses the cyanates by dioxygenase, converting them into ammonia and carbon dioxide (Knowles, 1976; Kunz et al., 1994).

Cyanide monoxygenase:

$$HCN + H_2O = HCONH_2 \tag{7.1}$$

$$HCN + O_2 + H^+ + NAD(P)H = HOCN + NAD(P)^+ + H_2O \tag{7.2}$$

Cyanide dioxygenase:

$$HCN + O_2 + 2H^+ + NAD(P)H = CO_2 + NH_3 + NAD(P)^+ \tag{7.3}$$

The presence of other by-products, formamide, and formate has also been found and revealed that when every factor is constant, a microbe will utilize a pathway most favoured by other factors such as the pH, availability of oxygen, carbon dioxide, and other source. A failed attempt to degrade cyanides with the washed cell of *Pseudomonas flourescens* under anaerobic conditions indicated the process being dependent on the oxygen of the system (Kunz et al., 1994).

7.7.2 Reductive Pathway

Although the anaerobic conditions for destructing cyanide are uncommon (Raybuck, 1992), numerous microbes follow this route to form ammonia and methane as the destructive products (Kao et al., 2003), shown in the following reactions:

$$HCN + 2H^+ + 2e^- = CH_2NH + H_2O + NAD(P)^+ = CH_2O \tag{7.4}$$

$$CH_2NH + 2H^+ + 2e^- = CH_3^-NH + 2H^+ + 2e^- = CH_4 + NH_3 \tag{7.5}$$

The production of formate as a by-product can lead to the hydrolysis of cyanides by methanogens that might be the dominant mode of cyanide degradation (Fallon et al., 1991). The use of isotopic carbon, $C^{14}N$, has not confirmed degradation via reduction and suggested the possibility of hydrolytic process (Raybuck, 1992).

7.7.3 Hydrolytic Pathway

Following the hydrolytic pathway, there is direct cleavage of the C≡N bond, eliminating the possibility of further reactivity (Raybuck, 1992). Cyanide loss even during cyanidation has been partly attributed to its hydrolytic decomposition (Adams, 1990). The decomposition can take place either through hydrolysis (catalysed by cyanidase forming by-products formic acid and ammonia) or through hydration (catalysed by cyanide hydratase forming the formamide by-product). The hydrolytic pathway is the most promising, with higher cyanide concentrations (Dumestre et al., 1997). Products from both hydrolysis and hydration are less toxic and can serve as food substrate for growing other microbes. The hydrolytic pathway requires only the presence of a functional enzyme (Raybuck, 1992; Dumestre et al., 1997) that can be an anaerobic or aerobic process.

Notably, the hydrolytic degradation of cyanides occurs either in a single step or two separate steps. In a two-step process, in the first step cyanide is catalysed by the enzyme (cyanide hydratase) forming an intermediate formamide, which is subsequently hydrolysed to formic acid (formate) and ammonia (as shown in Figure 7.7a); whereas, in a single-step process, cyanide is directly converted to formic acid and ammonia without any intermediate product formamide (as shown in Figure 7.7b).

7.7.4 Substitution/Addition Pathway

This degradation pathway exploits the thiophilic (a high affinity to sulphur) and the nucleophilic (a higher reactivity to the reaction sites of lower electron density) characteristics of cyanide ions (Raybuck, 1992). The degradation reactions require either sulphur transferase or a pyridoxal phosphate enzyme to convert cyanide into thiocyanates. Sulphurtransferases (EC 2.8.1.1–5) catalyse the transfer of sulphane sulphur from a donor molecule

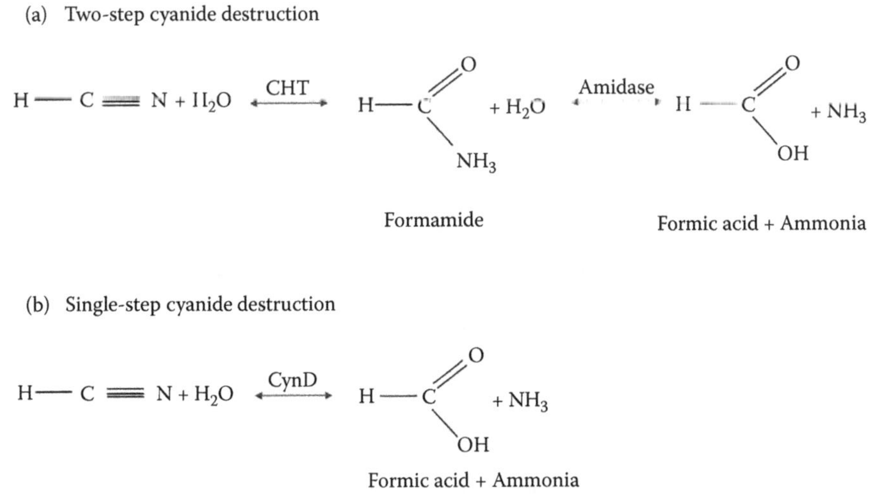

(a) Two-step cyanide destruction

Formamide   Formic acid + Ammonia

(b) Single-step cyanide destruction

Formic acid + Ammonia

FIGURE 7.7
A two- and single-step destruction process of cyanide using the microbial activity.

to a thiophilic acceptor. Rhodanase, 3-Mercaptopyruvate sulphurtransferase, and β-cyanoalanine enzymes have the major role in this pathway.

Rhodanase is a highly concentrated mitochondrial enzyme that detoxifies cyanide to convert it to thiocyanate (Raybuck, 1992). The degradation reaction occurs in two steps:

i. reduction of thiosulfate by a sulphur donation to the sulfhydryl of the protein to form a persulfide and a sulphite, and

ii. cyanide reacts with persulfide to produce less toxic thiocyanate and releases the thiol group enzyme molecule.

The reactions of both steps can be given as follows:

$$S_2O_3^{-2} + EnzS^- = EnzSS^- + SO_3^{-2} \tag{7.6}$$

$$CN^- + EnzSS^- = EnzS^- + SCN^- \tag{7.7}$$

Mercaptopyruvate is the single sulphur donor but there are many sulphur accepting thiophiles, including cyanide. The formation of an intermediate is not a prerequisite for enzymatic action in this pathway; rather, the substrates cyanide and 3-Mercaptopyruvate sulphurtransferase are simply bound in a ternary complex, and can be given as follows (Raybuck, 1992):

$$3-mercaptopyruvate + cyanide = pyruvate + thiocyanate \tag{7.8}$$

In this pathway of cyanide bio-degradation, substitution of a three-carbon amino acid with cyanide forms an intermediate β-cyanoalanine, which subsequently is hydrolysed to asparagine (Brysk et al., 1969; Brysk and Ressler, 1970; Rodgers and Knowles, 1978). The enzyme can be formed either from cysteine or the O-acetyl-serine, as shown in Figure 7.8a–c (Knowles, 1976).

7.7.5 Factors Affecting Bio-Degradation Process

It is well-known that microbes are pH sensitive. Bio-degradation of cyanide at a pH≥9 has been found to be extremely slow. It can be described by the unavailability of cyanide at high alkaline condition as it is strongly bonded to the complex rather than to a reduction in enzyme/bacterial activity (Dumestre et al., 1997). The optimal cyanide concentration in the solution to be treated is also important. The microbial activity is impeded at some elevated concentrations while in some extreme cases they cannot survive. The main factors of the process are aerobic or anaerobic and are discussed in the following paragraphs.

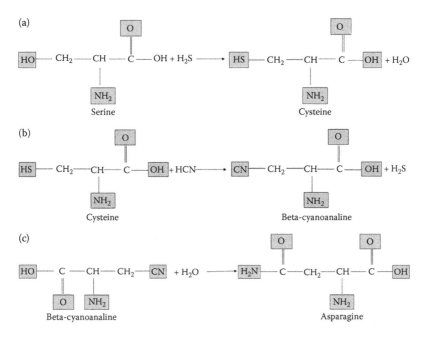

FIGURE 7.8
The substitution pathways for cyanide destruction by following the conversion of (a) serine to cysteine, (b) cysteine to β-cyanoalanine, and (c) hydrolysis of β-cyanoalanine to asparagine (adopted after modification from Knowles, 1976).

7.7.5.1 Aerobic Process

The aerobic process has received more attention than the anaerobic process over a half decade of time (Fallon, 1992). Under the aerobic condition, HCN is converted to hydrogen cyanate as shown in Equation 7.9. This process is convenient to deal with ~200 ppm CN in solution and requires an enzyme to catalyse the formation of hydrogen cyanate. Soon after, the hydrogen cyanate hydrolyses to form ammonia and carbon dioxide (as shown in Equations 7.9 and 7.10).

$$2HCN + O_2 + (enzyme) = 2HCNO \tag{7.9}$$

$$HCNO + H_2O = NH_3 + CO_2 \tag{7.10}$$

7.7.5.2 Anaerobic Process

Basically, the anaerobic process deals with free and complex cyanide degradation in the absence of air and/or oxygen, and takes place in highly reducing or reduced environments that are available in certain portions of tailing dams. Inhibition of anaerobes is somᵇow easy, due to their having a lot of metalloenzymes that can be easily targeted for inhibition by cyanide

(Cipollone et al., 2008). The presence of HS (at pH > 7) or H_2S (at pH < 7) is a prerequisite in an anaerobic degradation. The anaerobic degradation of cyanide can be represented as follows:

$$CN^- + H_2S = CNS^- + H_2 \qquad (7.11)$$

$$HCN + HS^- = CNS^- + H_2 \qquad (7.12)$$

The microbes capable of anaerobic degradation of cyanide are greatly sensitive to cyanide concentrations; less than 1 ppm CN can be lethal to methanogenic bacteria (Fallon et al., 1991; Cuzin and Labat, 1992). Nevertheless, the anaerobic processes do not receive much attention due to inefficient degradation of cyanide to a desired level which fits for its disposal to the environment, along with the undesirable product ammonia. Simply, as compared to 2 ppm cyanide threshold for anaerobic systems of bio-degradation, the aerobic systems have a threshold of >200 ppm cyanide (Smith and Mudder, 1991).

7.8 Environmental Impact, Limitations, and Challenges for the Microbial Activity in Gold Metallurgy

The microbial activity in gold metallurgy (at any stage of unit operation or in entire unit operations of oxidation, cyanidation, cyanide degradation, and adsorption) is an environmentally friendly operation. The bio-oxidation performed with microbes eliminates or lowers the uses of hazardous chemicals and rejects the use of energy intensive roasting/pressure oxidation process. In leaching of gold by the means of bio-cyanidation, regulation of cyanide concentration as low as 1 mM is significantly less when compared to the amount of cyanide in chemical cyanidation. This is quite good for handling for degradation of free cyanide and, it can be done again, by the microbial activity without using additional chemical reagents. Therefore, it is expected that in the coming years, new commercial plants will be commissioned to operate with the application of microbial activity, particularly in the mining industries.

Thus, more intensive research is needed to understand the microbial community distribution in order to disclose the mechanisms bind microbes-to-metals interaction from mineral ores, as well as the interactions between the mixed consortiums. Isolation of novel bacteria (mesophilic and thermophilic), archaea, or fungi or bioengineered species for the increased rate of sulphide removal and recovery of other precious and valuable metals during the treatment of refractory ores will be a major contribution to their potential field applications. A focused interdisciplinary collaboration among microbiologists, metallurgists, chemists, and engineers is needed to identify, characterize, select, and process development for an efficient recovery of gold.

Overall, the future goal of studies should pay attention to:

i. isolation and identification of potential oxidizing, reducing, and cyanidation bacteria from indigenous sources;

ii. heterotrophic bioleaching to extract gold from the ores;

iii. optimization of bio-oxidation, reduction, and electro-biochemical leaching to achieve higher selectivity with increased extraction efficiency; and

iv. cost estimation of the combinational processes for their commercial application.

References

Adams, M.D. 1990. The chemical baviour of cyanide in the extraction of gold. 1. Kinetics of cyanide loss in the presence and absence of activated carbon. *Journal of the Southern African Institute of Mining and Metallurgy.* 90(2): 37–44.

Adams, D.J., Komen, J.V., Pickett, T.M. 2001. Biological cyanide degradation. In: Young, C. (Ed.), *Cyanide: Social, Industrial and Economic Aspects.* The Metals Society, Warrendale, PA, pp. 203–213.

Ahmad, A., Senapati, S., Khan, M.I., Kumar, R., Ramani, R., Srinivas, V., Sastry, M. 2003. Intracellular synthesis of gold nanoparticles by a novel alkalotolerant actinomycete, Rhodococcus species. *Nanotechnology.* 14(7): 824.

Arrascue, M.L., Garcia, H.M., Horna, O., Guibal, E. 2003. Gold sorption on chitosan derivatives. *Hydrometallurgy.* 71(1): 191–200.

Askeland, R.A., Morrison, S.M. 1983. Cyanide production by *Pseudomonas fluorescens* and *Pseudomonas aeruginosa*. *Applied and Environmental Microbiology.* 45: 1802–1807.

Bosecker, K. 1997. Bioleaching: Metal solubilization by microorganisms. *FEMS Microbiology Reviews.* 20(3–4): 591–604.

Brandl, H., Bosshard, R., Wegmann, M. 2001. Computer-munching microbes: Metal leaching from electronic scrap by bacteria and fungi. *Hydrometallurgy.* 59(2): 319–326.

Brandl, H., Lᵎmann, S., Faramarzi, M.A., and Martinelli, D. 2008. Biomobilization of silver, gold, and platinum from solid materials by HCN-forming microorganisms, *Hydrometallurgy.* 94: 14–17.

Brysk, M.M., Ressler, C. 1970. γ-Cyano-α-l-aminobutyric acid a new product of cyanide fixation in *Chromobacterium violaceum*. *Journal of Biological Chemistry.* 245(5): 1156–1160.

Brysk, M.M., Lauinger, C., Ressler, C. 1969. Biosynthesis of cyanide from [2–14C-15N] glycine in *Chromobacterium violaceum*. *Biochimica et Biophysica Acta.* 184: 583–588.

Bulmer, C., Haas, D. 2000. Mechanism, regulation, and ecological role of bacterial cyanide biosynthesis. *Archives of Microbiology.* 173: 170–177.

Bunch, A.W., Knowles, C.J. 1982. Production of the secondary metabolite cyanide by extracts of *Chromobacterium violaceum*. *Journal of General Microbiology.* 128: 2675–2680.

Campbell, S.C., Olson, G.J., Clark, T.R., McFeters, G. 2001. Biogenic production of cyanide and its application to gold recovery. *Journal of Industrial Microbiology and Biotechnology.* 26: 134–139.

Castric, P.A. 1975. Hydrogen cyanide, a secondary metabolite of *Pseudomonas aeruginosa. Canadian Journal of Microbiology.* 21: 613–618.

Castric, P.A. 1977. Glycine metabolism by *Pseudomonas aeruginosa*: Hydrogen cyanide biosynthesis. *Journal of Bacteriology.* 130: 826–831.

Castric, P.A. 1981. The metabolism of hydrogen cyanide by bacteria. In: Vennesland, B., Conn, E.E., Knowles, C.J., Westley, J., Wissing, F. (Eds), *Cyanide in Biology.* Academic, London, pp. 233–261.

Chand, R., Wateri, T., Inoue, K., Kawakita, H., Luitel, H.N., Parajuli, D., Torikai, T., Yada, M. 2009. Selective adsorption of precious metals from hydrochloric acid solutions using porous carbon prepared from barley straw and rice husk. *Minerals Engineering.* 22: 1277–1282.

Cipollone, R., Ascenzi, P., Tomao, P., Imperi, F., Visca, P. 2008. Enzymatic detoxification of cyanide: clues from *Pseudomonas aeruginosa* Rhodanese. *Journal of Molecular Microbiology and Biotechnology.* 15(2–3): 199–211.

Crundwell, F.K. 2003. How do bacteria interact with minerals? *Hydrometallurgy.* 71: 75–81.

Cuzin, N., Labat, M. 1992. Reduction of cyanide levels during anaerobic digestion of cassava. *International Journal of Food Science & Technology.* 27(3): 329–336.

Das, T., Ayyappan, S., Chaudhury, G.R. 1999. Factors affecting bioleaching kinetics of sulphide ores using acidophilic micro-organisms. *BioMetals.* 12: 1–10.

Dew, D.W., Van Buren, C., McEwan, K., Bowker, C. 1999. Bioleaching base metal sulphide concentrates: A comparison of mesophile and thermophile bacterial cultures. In: Amils, R., Ballester, A. (Eds), *Biohydromatallurgy and the Environment Toward the Mining of the 21st Century, Part A.* Elsevier, Amsterdam, The Netherlands, pp. 229–238.

Dumestre, A., Chone, T., Portal, J., Gerard, M., Berthelin, J. 1997. Cyanide degradation under alkaline conditions by a strain of *Fusarium solani* isolated from contaminated soils. *Applied and Environmental Microbiology.* 63(7): 2729–2734.

Edwards, K.J., Bond, P.L., Gihring, T.M., Banfield, J.F. 2000. An archaeal iron-oxidizing extreme acidophile important in acid mine drainage. *Science.* 287: 1796–1799.

Erüst, C., Akcil, A., Gahan, C.S., Tuncuk, A., Deveci, H. 2013. Biohydrometallurgy of secondary metal resources: A potential alternative approach for metal recovery. *Journal of Chemical Technology and Biotechnology.* 88: 2115–2132.

Faramarzi, M.A., Brandl, H. 2006. Formation of water-soluble metal cyanide complexes from solid minerals by *Pseudomonas plecoglossicida. FEMS Microbiology Letters.* 259: 47–52.

Faramarzi, M.A., Stagars, M., Pensini, E., Krebs, W., Brandl, H. 2004. Metal solubilization from metal-containing solid materials by cyanogenic *Chromobacterium violaceum. Journal of Biotechnology.* 113: 321–326.

Fallon, R.D. 1992. Evidence of hydrolytic route for anaerobic cyanide degradation. *Applied and Environmental Microbiology.* 58(9): 3163–3164.

Fallon, R.D., Cooper, D.A., Speece, R., Henson, M. 1991. Anaerobic biodegradation of cyanide under methanogenic conditions. *Applied and Environmental Microbiology.* 57(6): 1656–1662.

Flaishman, M.A., Eyal, Z., Zilberstein, A., Voisard, C., Haas, D. 1996. Suppression of Septoria tritici blotch and leaf rust of wheat by recombinant cyanide producing strains of *Pseudomonas putida. Molecular Plant-Microbe Interaction.* 9: 642–645.

Fujiwara, K., Ramesh, A., Maki, T., Hasegawa, H., Ueda, K. 2007. Adsorption of platinum (IV), palladium (II) and gold (III) from aqueous solutions on l-lysine modified crosslinked chitosan resin. *Journal of Hazardous Materials.* 146: 39–50.

Gamez, G., Gardea-Torresdey, J.L., Tiemann, K.J., Parsons, J., Dokken, K., Yacaman, M.J. 2003. Recovery of gold (III) from multi-elemental solutions by alfalfa biomass. *Advances in Environmental Research.* 7(2): 563–571.

Golyshina, O.V., Pivovarova, T.A., Karavaiko, G.I., Kondrat'eva, T.F., Moore, E.R.B., Abraham, W.R., Lunsdorf, H., Timmis, K.N., Yakimov, M.M., Golyshin, P.N. 2000. *Ferroplasma acidiphilum* gen. nov., sp. nov., an acidophilic, autotrophic, ferrous-iron-oxidizing, cell-wall-lacking, mesophilic member of the Ferroplasmaceae fam. nov., comprising a distinct lineage of the Archaea. *International Journal of Systematic and Evolutionary Microbiology.* 50: 997–1006.

Greene, B., Hosea, M., McPherson, R., Henzl, M., Alexander, M.D., Darnall, D.W. 1986. Interaction of gold (I) and gold (III) complexes with algal biomass. *Environmental Science & Technology.* 20(6): 627–632.

Hagelüken, C. 2006. Recycling of electronic scrap at Umicore precious metals refining. *Acta Metallurgica Slovaca.* 12: 111–120.

Hol, A., van der Weijden, R.D., Weert, G.V., Kondos, P., Buisman, C.J.N. 2011. Processing of arsenopyritic gold concentrates by partial biooxidation followed by bioreduction. *Environmental Science and Technology.* 45: 6316–6321.

Hoque, M.E., Philip, O.J. 2011. Biotechnological recovery of heavy metals from secondary sources—An overview. *Materials Science and Engineering: C.* 31(2): 57–66.

Ilyas, S., Ruan, C., Bhatti, H.N., Ghauri, M.A., Anwar, M.A. 2010. Column bioleaching of metals from electronic scrap. *Hydrometallurgy.* 101: 135–140.

Ilyas, S., Lee, J-c. 2014. Biometallurgical recovery of metals from waste electrical and electronic equipment: A review. *ChemBioEng Reviews.* 1(4): 148–169.

Ilyas, S., Lee, J., Kim, B.-S. 2014. Bioremoval of heavy metals from recycling industry electronic waste by a consortium of moderate thermophiles: Process development and optimization. *Journal of Cleaner Production.* 70: 194–202.

Ilyas, S., Lee, J. 2015. Bioprocessing of electronic scraps. In: Abhilash, Pandey, B.D., Natrajan, K.A. (Eds), *Microbiology for Minerals, Metals, Materials and the Environment.* CRC Press, Boca Raton, FL, pp. 307–328.

Ilyas, S., Kim, M-s., Lee, J-c. 2017a. Integration of microbial and chemical processing for a sustainable metallurgy. *Journal of Chemical Technology and Biotechnology.* doi:10.1002/jctb.5402.

Ilyas, S., Kim, M-s., Lee, J-c., Jabeen, A., Bhatti, H.N. 2017b. Bio-reclamation of strategic and energy critical metals from secondary resources. *Metals.* 7(207): 1–7.

Ishikawa, S.I., Suyama, K., Arihara, K., Itoh, M. 2002. Uptake and recovery of gold ions from electroplating wastes using eggshell membrane. *Bioresource Technology.* 81(3): 201–206.

Kao, C.M., Liu, J.K., Lou, H.R., Lin, C.S., Chen, S.C. 2003. Biotransformation of cyanide to methane and ammonia by *Klebsiella oxytoca. Chemosphere.* 50(8): 1055–1061.

Khoo, K.M., Ting, Y.P. 2001. Biosorption of gold by immobilized fungal biomass. *Biochemical Engineering Journal.* 8(1): 51–59.

Kita, Y., Nishikawa, H., Takemoto, T. 2006. Effects of cyanide and dissolved oxygen concentration on biological Au recovery. *Journal of Biotechnology.* 124: 545–551.

Knowles, C.J. 1976. Microorganisms and cyanide. *Bacteriological Reviews.* 40(3): 652.

Kulpa, C.F. Brierley, J.A. 1993. Microbial deactivation of preg-robbing carbon in gold ore. In: Torma, A.E., Wey, J.E., Lakshmanan, V.L. (Eds), *Biohydrometallurgical Technologies, Vol. 1*. The Minerals, Metals and Materials Society, Warrendale, PA, pp. 427–435.

Kunz, D.A., Wang, C.S., Chen, J.L. 1994. Alternative routes of enzymic cyanide metabolism in Pseudomonas fluorescens NCIMB 11764. *Microbiology*. 140(7): 1705–1712.

Kuyucak, N., Volesky, B. 1988. Biosorbents for recovery of metals from industrial solutions. *Biotechnology Letters*. 10(2): 137–142.

Laville, J., Blumer, C., Schroetter, C.V., Gaia, V., D'efago, G., Keel, C., Haas, D. 1998. Characterization of the hcnABC gene cluster encoding hydrogen cyanide synthase and anaerobic regulation by ANR in the strictly aerobic biocontrol agent *Pseudomonas fluorescens* CHAO. *Journal of Bacteriology*. 180: 3187–3196.

Lawson, E.N., Barkhuizen, M., Dew, D.W. 1999. Gold solubilisation by the cyanide producing bacteria *Chromobacterium violaceum*. In: Amils, R., Ballester, A. (Eds), *Biohydrometallurgy and the Environment toward the Mining of the 21st Century*. Elsevier, New York, pp. 239–246.

Lee, J.-C., Srivastava, R.R. 2016. Leaching of gold from the spent/end-of-life mobile phone-PCBs. In: Sabir, S. (Ed.) *The Recovery of Gold from Secondary Resources*. Imperial College Press, London, UK, pp: 7–56.

Lin, Z., Wu, J., Xue, R., Yang, Y. 2005. Spectroscopic characterization of Au^{3+} biosorption by waste biomass of *Saccharomyces cerevisiae*. *Spectrochimica Acta*. 61: 761–765.

López, M.L., Parsons, J.G., Videa, J.R.P., Gardea-Torresdey, J.L. 2005. An XAS study of the binding and reduction of Au(III) by hop biomass. *Microchemical Journal*. 81: 50–56.

Maruyama, T., Matsushita, H., Shimada, Y., Kamata, I., Hanaki, M., Sonokawa, S., Kamiya, N., Goto, M. 2007. Proteins and protein-rich biomass as environmentally friendly adsorbents selective for precious metal ions. *Environmental Science and Technology*. 41: 1359–1364.

Merchant, B. 1998. Gold, the noble metal and the paradoxes of its toxicology. *Biologicals*. 26: 49–59.

Michaels, R., Hankes, L.V., Corpe, W.A. 1965. Cyanide formation by nonproliferating cells of *Chromobacterium violaceum*. *Archives of Biochemistry and Biophysics*. 111: 121–125.

Mousavi, S.M., Yaghmaei, S., Vossoughi, M., Jafari, A., Hoseini, S.A. 2005. Comparison of bioleaching ability of two native mesophilic and thermophilic bacteria on copper recovery from chalcopyrite concentrate in an airlift bioreactor. *Hydrometallurgy*. 80: 139–144.

Mullen, M.D., Wolf, D.C., Ferris, F.G., Beveridge, T.J., Flemming, C.A., Bailey, G.W. 1989. Bacterial sorption of heavy metals. *Applied and Environmental Microbiology*. 55(12): 3143–3149.

Niu, H., Volesky, B. 1999. Characteristics of gold biosorption from cyanide solution. *Journal of Chemical Technolology and Biotechnology*. 74: 778–784.

Ofori-Sarpong, G., Tien, M., Osseo-Asare, K. 2010. Myco-hydrometallurgy: Coal model for potential reduction of preg-robbing capacity of carbonaceous gold ores using the fungus, Phanerochaete chrysosporium. *Hydrometallurgy*. 102: 66–72.

Ofori-Sarpong, G., Osseo-Asare, K., Tien, M. 2011. Fungal pretreatment of sulphides in refractory gold ores. *Minerals Engineering*. 24: 499–504.

Olson, G.J. 1994. Microbial oxidation of gold ores and gold bioleaching. *FEMS Microbiology Letters*. 119: 1–6.

Parajuli, D., Kawakita, H., Inoue, K., Ohto, K., Kajiyama, K. 2007. Persimmon peel gel for the selective recovery of gold. *Hydrometallurgy*. 87: 133–139.

Paterson, C.J. 1990. Ore deposits of gold and silver. In: Arbriter, A., Han, K.N. (Eds), *Gold: Advances in Precious Metals Recovery*. Gordon and Breach, New York, Ch. 1, pp. 49–116.

Pethkar, A.V., Paknikar, K.M. 1998. Recovery of gold from solutions using *Cladosporium cladosporioides* biomass beads. *Journal of Biotechnology*. 63(2): 121–136.

Rawlings, D.E. 2005. Characteristics and adaptability of iron- and sulphur-oxidizing microorganisms used for the recovery of metals from minerals and their concentrates. *Microbial Cell Factories*. 4: 13–16.

Raybuck, S.A. 1992. Microbes and microbial enzymes for cyanide degradation. *Biodegradation*. 3: 3–18.

Rodgers, P.B., Knowles, C.J. 1978. Cyanide production and degradation during growth of *Chromobacterium violaceum*. *Journal of General Microbiology*. 108: 261–267.

Rohwerder, T., Gȯrke, T., Kinzler, K., Sand, W. 2003. Bioleaching review part A: Progress in bioleaching: Fundamentals and mechanisms of bacterial metal sulphide oxidation. *Applied Microbiology and Biotechnology*. 63(3): 239–248.

Romero-González, M.E., Williams, C.J., Gardiner, P.H.E., Gurman, S.J., Habesh, S. 2003. Spectroscopic studies of the biosorption of gold (III) by dealginated seaweed waste. *Environmental Science and Technology*. 37: 4163–4169.

Sand, W., Gȯrke, T., Hallmann, R. 1995. Sulfur chemistry, biofilm, and the (in) direct attack mechanism a critical evaluation of bacterial leaching. *Applied Microbiology and Biotechnology*. 43: 961–966.

Smith, A., Mudder, T. 1991. *Chemistry and Treatment of Cyanidation Wastes*. Mining Journal Books Ltd., London, p. 345.

Soleimani, M., Kaghazchi, T. 2008. Adsorption of gold ions from industrial wastewater using activated carbon derived from hard shell of apricot stones: An agricultural waste. *Bioresource Technology*. 99(13): 5374–5383.

Tasdelen, C., Aktas, S., Acma, E., Guvenilir, Y. 2009. Gold recovery from dilute gold solutions using DEAE-cellulose. *Hydrometallurgy*. 96: 253–257.

Torres, E., Mata, Y.N., Bl'azquez, M.L. 2005. Gold and silver uptake and nanoprecipitation on calcium alginate beads. *Langmuir*. 21: 7951–7958.

Townsley, C.C., Ross, I.S. 1986. Copper uptake in *Aspergillus niger* during batch growth and in non-growing mycelial suspension. *Experimental Mycology*. 10: 281–288.

Tributsch, H. 2001. Direct versus indirect bioleaching. *Hydrometallurgy*. 59: 177–185.

Tsuruta, T. 2004. Biosrption and recycling of gold using various microorganisms. *Journal of General and Applied Microbiology*. 50: 221–228.

Watling, H.R. 2006. The bioleaching of sulphide minerals with emphasis on copper sulphides: A review. *Hydrometallurgy*. 84: 81–108.

Wissing, F. 1974. Cyanide formation from oxidation of glycine by a *Pseudomonas* species. *Journal of Bacteriology*. 117: 1289–1294.

Wissing, F. 1975. Cyanide production from glycine by a homogenate from a *Pseudomonas* species. *Journal of Bacteriology*. 121: 695–699.

Wissing, F., Andersen, K.A. 1981. The enzymology of cyanide production from glycine by a *Pseudomonas* species. In: Vennesland, B., Conn, E.E., Knowles, C.J., Westley, J., Wissing, F. (Eds), *Cyanide in Biology*. Academic, London, pp. 275–288.

Yen, W.T., Amankwah, R.K., Choi, Y. 2008. Microbial pre-treatment of double refractory gold ores. In: Young, C.A., Taylor, P.R., Anderson, C.G., Choi, Y. (Eds), *Proceedings of the 6th International Symposium, Hydrometallurgy. 2008*. SME, Littleton, CO, pp. 506–510.

8

Human Perspectives on Gold Exploitation and Case Studies

Sadia Ilyas*, Shafaq Masud*, and Jae-chun Lee[†]

8.1 Introduction

As previously described, tedious efforts are required to extract gold, which is universally considered to be a symbol of prosperity and wealth. Nevertheless, the human costs of mining, metallurgy, conscripted, and convict labour with their unhealthy degraded living conditions cannot be detached from the enduring value associated with gold. As the world is becoming more concerned with sustainability rather than only the efficient extraction of this precious metal (which is sometimes referred to as "dirty gold"), it is imperative to know the human perceptions on gold metallurgy. To clarify, the human perceptions are based on the practices of the gold mining and metallurgical industries, including deep underground processes and surface working (amalgamation) processes and their impacts on the society and environment. Most of the adversity caused by mercury used in the amalgamation of gold has been discussed in Chapter 2; therefore the environmental hazards of mercury are not included in this chapter.

8.2 Vitality of Human Perception on the Gold Industry

The importance of this topic can be understood by a recent study carried out on global perceptions of the gold mining industry by the World Gold Council to reveal the challenges the industry faces now and in

* Mineral and Material Chemistry Lab, Department of Chemistry, University of Agriculture Faisalabad, Pakistan.
† Minerals Resources Research Division, Korea Institute of Geoscience and Mineral Resources, Daejeon, South Korea.

the future (World Gold Council Report, 2013). The results obtained by the study are shown in Figure 8.1 and clearly indicate that at present time the community issues to obtain a social license to operate is the most challenging factor. It is followed by the factor of human health and environmental issues. However, the competition for natural resources (including water) is identified as the third most important factor at the present time, but will be the greatest challenge in the next 20 years. Therefore, it is of vital importance for any company that wants to ensure the reserves of resources in the future, to obtain a social license to operate as soon as possible by solving the issues related to human health and the environment. In this context, all three factors are interlinked and of equal importance. Furthermore, the report reveals the need for industry in the next 10 years to be an eco-friendly (Figure 8.2), respectful engagement with local communities and address the issues related to artisanal and small-scale mining (World Gold Council Report, 2013). However, the importance of artisanal mining for providing livelihood opportunities to many peoples cannot be neglected. Thus, the perceptions for artisanal and small-scale mining have both positive and negative aspects. Approximately 63% of stakeholders from the study accepted the negative

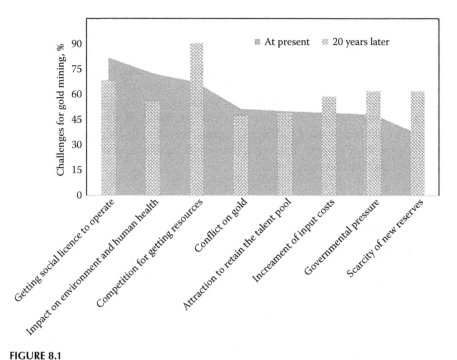

FIGURE 8.1
The present and future challenges for gold mining from the human perspective. (Modified and Adopted from the World Gold Council Report, 2013)

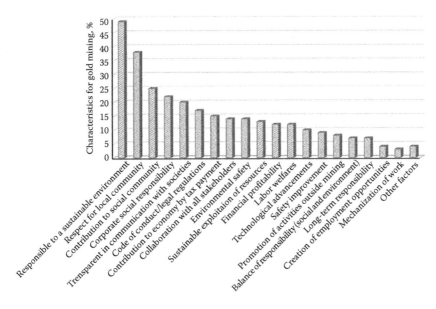

FIGURE 8.2
The priorities of the characteristics that are required for gold mining in the next 10 years of operation. (Modified and Adopted from the World Gold Council Report, 2013)

impacts of artisanal mining as worse for the livelihood of local communities, but 65% of stakeholders have also agreed to view artisanal mining as an opportunity for those people who do not have other options. Throughout the Brazilian Amazon, more than 65,000 small-scale miners are producing 90% of Brazil's total gold production capacity and discharging ~120 t mercury annually to the environment.

The environmental degradation causing the health risk of the local population is one of the major factors creating a direct perception on any issue, and it is the same in the case of gold mining and metallurgy, which will be discussed first.

8.3 Health Risks at Localities of Underground Mining

A decreased life expectancy; increased chances of cancers of the bronchus, stomach, liver, and trachea; increased frequency of pleural diseases and pulmonary tuberculosis; insect borne diseases; hearing loss; skin diseases; and viral and bacterial diseases are often found in the workers and locality of underground mining. Eisler (2003) has widely reported the documented problems of peoples from North and South America, Australia, and South Africa.

The Australian gold miners in tropical regions are impotent against vector-borne diseases, dengue, and malaria due to larvae and pupae breeding of *Aedes aegypti* in flooded unused mining shafts. The Western Australia miners are found to suffer lung cancer, silicosis, and pulmonary tuberculosis due to high dust exposure during mining activity. The gold miners of Canada and the United States have found significantly higher rates of lung cancer, silicosis, and pulmonary tuberculosis. Gold miners with more than 5 years of working in Ontario and more than 20 years of working in North American mines showed increased risks for primary cancers. The health check-up of the workers of underground mining in a South Dakota mine has shown a 68%–84% increased exposure to silica and non-asbestiform materials, resulting in a 13% higher rate of lung cancer than the common people of the United States. Several deaths in Colombian mines have been reported while digging at a condemned mine by ignoring the government's warning. The prevalence of malaria has increased remarkably in Brazil with more than 6,000,000 cases. The attacks of vampire bats, *Desmodus rotundus* have been recorded in southeastern Venezuela, possibly due to the loss of normal habitat of bats from mine sites. In northeastern Gabon (Africa), the miners practicing gold panning are found to be affected by leptospirosis and Ebola virus causing lethal haemorrhagic fever. Noise pollution laws are usually not enforced in developing countries; hence, 20% of miners experienced hearing loss in central Ghana. South African miners have the highest rate of tuberculosis worldwide. Notably, the deaths caused by cancers (liver, oesophageal, and lung) in African miners working underground showed a higher rate. The (white) South African miners who spent ~85% working life in gold mines have a 30% chance of dying sooner than others.

8.4 Health Risks at Localities of Surface Mining

A range of risks identified with surface mining activities (including the ergonomic stresses; noise; unspecified lifestyle factors; limited exits in underground tunnels; traumatic injuries; and a lack of safety culture, training, or personal protective equipment) have been summarized in Table 8.1. Miners involved in surface mining activities which use elemental mercury to amalgamate and extract gold have been found to be heavily contaminated with mercury. Among individuals occupationally exposed, the contamination of mercury in drinking water, air, fish diet, blood, urine, hair, and other tissues significantly exceeds the exposure limits proposed by various regulatory bodies in the Philippines (Appleton et al., 1999; Akagi et al., 2000), Kenya (Ogola et al., 2002), and Brazil (Malm et al., 1990) for protecting the human health. Several cases in these countries have been reported in which many persons died of mercury contamination. The top health and safety risks of surface mining are summarized in Table 8.2.

TABLE 8.1

The Prominent Health and Safety Risks to Miners Involved in Galamsey or
Artisanal Gold Mining

Risks to Miners	Number of Respondents	%
Mercury	5	63
Pit collapses and rock falling	4	50
Ventilation and respiratory diseases	4	50
Environment and community pathogens	2	25
Ergonomic stresses	1	13
Other lifestyle factors	1	13
Noise issue	1	13
Restricted/limited exits	1	13
Injuries by trauma	1	13
Lack of safety training and protective equipment	1	13
Temporary living conditions	1	13

TABLE 8.2

The Prominent Health and Safety Risks to the Community and Family
of Miners Involved in Galamsey or Artisanal Gold Mining

Risks to Miners	Number of Respondents	%
Water contamination	5	63
Mercury	4	50
Diseases caused by infections	3	38
Exposure to cyanide	2	25
Child labour	2	25
Other lifestyle factors	1	13
Domestic/gender violence	1	13
Rock falling/landslides	1	13
Abandoned mines hazards	1	13
Conflict between locals and migrants	1	13

Miners are exposed not only to mercury in the air, water, and fish but also
suffer from mercury contamination in tissues due to exposure to mercury
vapor, methylmercury, and inorganic mercury. The increased mercury levels
in urine are positively correlated with increased consumption of fish and
alcohol, working hours per day, and dental amalgam fillings (Santa Rosa
et al., 2000). In China, the mean concentration of mercury in urine from a
gold mining community has been found to be up to 540 µg/L, which is many
folds higher than the accepted limits. Methylmercury concentration in hair
should be limited to 4–7 mg/kg; a higher concentration of 10–20 mg/kg can
cause abnormal infant development, and 50–100 mg/kg can cause paraes-
thesia (de Lacerda and Salomons, 1998). Among the Brazilian gold miners,

TABLE 8.3

The Challenges to Mitigate the Health and Safety Risks to Miners
Involved in Galamsey or Artisanal Gold Mining

Risks to Miners	Average Respondent
Negligence of policy and regulatory frameworks	2
Illegal mining	3
Lack of training	4
Perceived tradeoff between health and safety	4
Inadequate training of the enforcement personnel	5
Lack of funds for materials and equipments	5
Inadequate risk awareness	5
Resistance to intervention by government or NGOs	6

those who consume more than 100 g/day of fish need to have their levels of methylmercury monitored (Krig et al., 1997). Elevated concentrations (25–37 mg/kg) of total mercury are found in the analysed hair of villagers near gold mining areas; >90% of mercury in hair is in the form of methylmercury (Akagi et al., 1995). Mercury concentration in breast milk from nursing mothers in the Brazil Amazon Basin was found to be up to 24.8 μg/kg (an average of 5.8 μg/kg against 0.9 μg/kg in the United States), a level that is 53% higher than the limits designated by the World Health Organization for infant consumption. The mercury in breast milk is not significantly correlated to the mercury contamination in hair (Barbosa et al., 1998). Inhalation of metallic mercury vapor in ambient air of gold dealer shops and workplaces in gold mining areas exceeds 1664 μg/m^3 in the Philippines, 183 μg/m^3 in Venezuela, and 292 μg/m^3 in Brazil, compared to the recommended exposure level of 50 μg Hg/m^3 (de Lacerda and Salomons, 1998). Approximately 72% of Philippine workers in amalgamation are classified as mercury-intoxicated (Drasch et al., 2001). In Jiangxi Province of China, approximately 200 small-scale gold mines use mercury amalgamation to extract gold, usually conducted in private residences with excessive mercury contamination up to 2600 μg/m^3 in workrooms and up to 1000 μg/m^3 at workshops; these were later prohibited by imposing China's national environmental legislation (Lin et al., 1997). The challenges in mitigating health and safety risks associated with surface mining are summarized in Table 8.3.

8.5 Case Study on Artisanal and Small-Scale Gold Mining

A case study has been carried out on a Canadian mining company, Golden Star Resources, at the operational sites at Bogoso, Prestea and Wassa in Ghana. The following are positive and negative impacts of large-scale mining on the economy and local communities; however, more detailed information can be obtained in the literature reported by Akabzaa, (2000, 2001), Akabzaa and Darimani (2001), Ayine (2001), Hilson (2002), Akpalu and Parks (2007), Ghana Chamber of Mines (2007), CHRAJ (2008), Garvin et al. (2009), Minerals Commission of Ghana, and Nyame et al. (2009).

8.5.1 Positive Effects

The liberalization in the mining sector through privatization and the introduction of comprensive regulations (Minerals and Mining Law, 1986/ PNDCL 153) resulted in flooded foreign direct investment in mining. This translated into increased opportunities for employment (Akabzaa and Darimani, 2001), including a significant number of indirect employment opportunities given by the mining support companies (e.g. assay laboratories, equipment leasing and sales agencies, and catering and security catering agencies).

The mining sector established itself as the single largest foreign exchange earner by contributing ~45% of the total earning of Ghana via foreign exchanges. The large-scale mining sector accounts for 7%–9% of total revenues, which equals $43,226,713 for the government. Additionally, various districts and municipal offices also earned 7.59 billion Ghana cedi in 2007 in the form of several taxes (Ghana Chamber of Mines, 2007). The identified "social multipliers" directly upgrading the rural communities involved in the mining sector can be itemized as follows:

- Banking: The mining operations at Tarkwa, Obuasi, and Prestea have encouraged the banking sector to set up local branches, thereby generating local employment to provide services to community-based businesses.
- Communication: The need for viable communication services facilitated telephone and mobile phone services for those living around the mines.
- Electricity: The mining companies have provided electricity and electrical equipment to facilitate power to mining communities.
- Infrastructure development: The construction of access roads to rural communities has stimulated economic growth and has improved connectivity to the necessary services.
- Education and Health: Many schools and hospitals have been built by mining companies to support the basic needs of the local communities, which are typically in under-serviced rural areas.
- Human resource development: To run their mining operations, the mining companies require trained manpower and have trained many people to enhance their skills; the mining companies later recruited these people to work in the mines.

8.5.2 Negative Effects

Although there are positive effects on the economy and employment generation caused by the flow of investments from the gold mining companies, a
range of social and environmental concerns have arisen as well in Tarkwa and
Bogoso, Prestea. The rush to gold has resulted in moving the mining from
underground to open pit operations that employ fewer people for mining the
ores. The switching from less profitable underground mining to the more
profitable surface mining operations generated social conflict, as the unprofitable mines closed due to the substantial lay-offs. The need for more land
to develop open pit mines than required for traditional underground mines
could result in decreasing operational areas from the land base and also
decreasing the agriculture land (Akabzaa, 2000). When mining was liberal-
ized in Ghana, the new democracy of the country in 1992 did not immediately
translate to actual protection for communities adversely affected by mining.
Until 1994, no stringent act of environmental protection was in the country
and was only regulated by the Environmental Protection Council (1974), which
was not independent and was working under the political until 1999 when the
Golden Star Resources entered mining in Ghana by purchasing the Bogoso
mines. The increase in foreign mining companies engaged in surface mining
coincided with allegations of human rights abuses; in reaction to local communities trying to protect the environment from degradation, the state security
personnel harassed and forcibly moved the members of local communities
opposing new mine operations (Ghanaian Chronicle, 1998). The following
concerns associated with new mine development in Ghana have been found:

- Environmental issues include water and air pollution; loss and degradation of agriculture land; and pollutions by the means of air,
  water, and noise. High rainfall poses problems for tailings disposal
  facilities and the control of run-off from mining areas is required,
  otherwise disrupting the availability of drinking water. Mining at
  Western Ghana typically entails heap leaching and/or carbon-in-
  leach processes using cyanide as the lixiviant that occasionally spills
  and contaminates the surface and ground waters.

- Open-pit mining is noisy and generates dust; therefore, local communities are relocated to reduce disturbance. Depending on their
  proximity to the source, they can be affected by the emissions.
  Children are especially vulnerable to dust. Dust settled on plants
  reduces vegetation and crops, whereas the noise and vibrations from
  blasting affect the safety to the buildings close to operations.

- The changing of less profitable underground mining to profitable
  surface mining operations caused the closure of the unprofitable
  mines, which resulted in unemployment for many workers in mining areas of Tarkwa, Prestea, and Obuasi. The new open pit mines
  hired for new jobs but in fewer numbers than workers who were

unemployed from the closing of underground mining. Moreover, recent advances in both underground and open-pit mining require more skilled workers. Job loss and inability to find new unskilled jobs with similar pay causes migration of people to other areas in hopes of finding employment with the mining companies, which further exacerbates the unemployment situation.

- The rural areas of Western Ghana are highly involved in informal employment (Amenumey, 2008; Aryeetey and Kanbur, 2008); the majority of people are farmers. People who are generally poorly educated can be considered employed, even if not counted as part of the labour force. The growth in surface mining has meant a greater disturbance of the area by the means of inappropriate compensation. To determine compensation the 2006 Minerals and Mining Act (Act 703) was introduced. But concerns of the ability of farmers to represent their interests have yet to be addressed by the government.

- The illegal small-scale mining in Ghana (termed as *galamsey*) includes some within the legal mining concessions. Some license holder miners do not actually mine their own concessions, work elsewhere as "illegal" miners, and use their licenses to cover up the illegal activity. Despite labour laws, these activities are dangerous and often use child labour, violating human rights; the government also loses taxes and revenues.

8.5.3 Multi-Perspectives on Issues Pertaining to Mining

An economic recovery program in the early 1980s, revision of mining laws in 1986, and the Mining Act of 2006 are some government initiatives taken to attract foreign direct investment to the mining sector of Ghana and earn a royalty from it. The rights of land owners are defined in the Mining Act of 2006 (Sections 72 75), together with dispute mechanisms include the option to take grievances to court. The anti-mining NGOs have a general consensus that mining companies co-opted key societal players and b›aved aggressively with impunity towards local communities. The increased presence of law enforcement agencies, shortage of farming land, and encroachment of surface mines into forest reserves is a major concern voiced by NGOs. NGOs engage in direct community activism along with se–ing to influence the national mining policy. The compensation regime for destroyed crops is deemed by the NGOs to be woefully inadequate, especially where the cash crop cocoa is concerned, as the compensation for deprivation of use is negotiated between the company and farmers. The Centre for Public Interest Law represents those persons who wish to take their grievances to the court. A key dimension of such a national strategy is for the government to provide the means for small-scale mining to operate viably alongside, but not in competition with, large-scale mining.

By showing that the mining industry generates significant revenue for the government of Ghana (it contributes 5% of GDP and 10% of Ghana's revenue inflows), miners put concerns over the social responsibility b›ind basic infrastructure. A disproportionate use of revenue derived from mining activities is often imposed and grievances for the responsibilities of central and local government to provide transport and social services have been asked by the miners. A serious omission of human rights by the galamsey operators who carry out armed attacks against employees of mining companies has been reported as a serious issue. The mining sector is recognizing the impetus of corporate social responsibility and policing them to learn from each other by adopting standard practices which are sensitive to the community and address the needs of local economic development. The community perspectives include the sense of injustice when mining operations dislocate the local economy in favour of migrated people, but illegal miners engaged with local people in their trade focusing on matters like health and environment create doubt on the imposed social mitigation initiatives and prefer the community led projects.

The influx of foreigners, predominantly from China, has accelerated the galamsey although Ghanaian law prohibits small-scale mining by any foreign persons. The Chinese have invested a substantial amount into mining with an elevated rate of ore extraction and environmental damages from the use of heavy machinery. Approximately 50,000 "gold se–ers" from China, exclusively from the Guangxi province, are settled in rural areas creating Chinese mining communities (Burrow and Bird, 2017). Over 300 Chinese migrant workers are allegedly employed at each galamsey site together with a number of Ghanaians, making the Ghanaians feel hostile toward the outsiders. The reason is that at the start of April 2017 the government of Ghana issued a 3-we– ultimatum to stop galamsey in the country or face prosecution. Som›ow it worked and reports say that hundreds of excavators used in artisanal gold mining/galamsey were voluntarily removed from mine sites. It showed that the problem of galamsey in Ghana is very difficult to solve but taking serious steps with a wide range of compr›ensively pursued strategies can work out positively.

With a fair contribution to the national economy of Burkina Faso, about 700,000 families are dependent on the artisanal gold mining (ASGM) activities. The perception on ASGM evaluated by Sana et al. (2017) has revealed the increased awareness among local miners on the issues related to the environment and public health. A cross-sectional survey conducted in artisanal mines located at Bouda and Nagsene revealed that all the people residing in nearby localities are aware of the environmental and health impacts. More than 60% of miners identified 3 out of 5 adverse health effects and 49.5% understood the environmental impacts caused by the ASGM.

Myanmar offers peculiar opportunities for foreign direct investment and biodiversity conservation as well, by linking the projected economic growth with the exploitation of mineral resources (including gold bearing ores).

In a recent study carried out by Papworth et al. (2017), new land uses have been found to be introduced through the mining development, but with unknown repercussions for local people and biodiversity conservation. In an area of ~21,800 km² of Myanmar's Hukaung Valley, local communities are getting benefit from the work and trade opportunities offered by gold mining; however, they continue to be dependent on forests to fulfil other needs like materials for house construction and food items. On the other hand, the gold mining concessions cause the potential reduction of tree cover and forest resources, which further marginalizes the local people.

8.6 Case Study on Gold Cyanidation

Cyanide has a toxicity limit <500 ppb HCN and 10 ppm as weak acid dissociable cyanide found in the discharge. It is the major method used for gold extraction, and has caused people from various part of the world to suffer from time-to-time (Table 8.4). The residues after cyanidation are usually stored in tailings dams for several years, especially if no cyanide destruction step was followed. For example, in the year 2000 at Baia Mare (Romania), a dam containing cyanide tailings burst. Approximately 100,000 m³ of cyanide and heavy metals spilled into the Lupus stream, poisoning the water of the Szamos, Tisza, and Danube Rivers, killing several hundred tonnes of fish, and contaminating the drinking water in Slovakia and Hungary.

TABLE 8.4

Some Accidents Which Caused Serious Cyanide Contamination

Accident Location	Year	Impact
Stava, Italy	1985	Cyanide tailings flowed up to 8 km area causing the loss of 269 lives
Northparkes mine, NSW, Australia	1995	Thousands of non-migratory water birds were killed due to a poor understanding of cyanidation, and inappropriate analytical procedures (Environment Australia, 2003)
Barskaun River, Kyrgyzstan	1998	~1800 kg NaCN contaminated the Barskaun River following a truck accident enroute to Kumtor mine (Hynes et al., 1999)
Baia Mare, Romania	2000	Breached tailing impoundment released a cyanide plume and killed fishes in the Tisza and Danube Rivers (UNEP/OCHA, 2000; Environment Australia 2003)
Wassa, Ghana	2001	Cyanide contaminated water from Tarkwa gold mines entered the Asuman River, killing fish and disrupting local water supplies. Another discharge from a ventilation shaft in 2003 r–indled community health and safety concerns
San Andres mine, Honduras	2002	A misoperation caused a number of valves to open in the cyanide plant releasing 1200 L cyanide into the Lara River

As a result of such accidents, the International Cyanide Management Code formed as a voluntary initiative whose recommendations are applied in the European Union countries. A number of countries like the Czech Republic and Germany (in 2002) and Hungary (in 2009) have banned cyanide.

The study indicates that spilling of cyanide is usually caused by the following sources, which can be designed or regulated to avoid accidents: (i) transportation of cyanide to the site for use (~14% cases), (ii) tailing dam failure (~72% cases), (iii) seepage via the heap/dump failure, and (iv) clandestine usage by artisanal miners. Three reasons are mainly responsible for occasional spilling out of cyanide:

i. Lack of water balance and management,

ii. Implementation of improper water treatment capabilities, and

iii. Absence of integrity and secondary containment infrastructures.

It is noteworthy that the real environmental threat that occurs during dam failure is not from the cyanide but due to the release of toxic and heavy metals like cadmium, lead, mercury, copper, arsenic, zinc, etc.

A case study about the Newmont Waihi Gold Mine located 150 km southeast of Auckland (New Zealand) with ~4700 people residing nearby presents a range of communication processes initiated by the mining authority with the purpose of maintaining positive relationships with key stakvolders and discussing matters of mutual concern. One topic discussed was the safe management of cyanide transportation, storage, and its uses. As a result, links between the mining company and education institutions have grown, the Martha Mine became a popular destination for a variety of educational groups, and gold mining is a part of the secondary school syllabus to teach the gold mining as centre-of-interest topics as part of the technological curriculum. Educational tours for the students of universities and polytechnics are also arranged. The mine has also made other efforts towards improving community engagement, sustainable transportation of cyanide and its delivery, process improvements to reduce weak acid dissociable (WAD) cyanide concentrations in tailing discharge, and cyanide destruction methodologies.

8.7 Case Study on Acid Mine Drainage from Gold Mining

As time passed, gold mining activities have gone deeper and pyrite bearing gold ores have been encountered, hindering the leaching process as described in earlier chapters. Ore mined underground brought for either amalgamation or cyanidation after fine milling generates a huge number of tailings sent to the dump, typically containing 0.5 g/t gold and a significant amount of unbroken pyrites which are subjected to slow atmospheric oxidation

(Naicker et al., 2003). Many of such tailing dumps in the Johannesburg area have remained undisturbed for almost a century, and the long exposure to oxygenated rainwater caused the oxidation of sulphides up to 5 m depths (Marsden, 1986). Notably, the oxidation by ferric iron can be two folds higher than oxidation carried out by oxygen (Chandra and Gerson, 2010); nevertheless, the oxidation of Fe^{2+} to Fe^{3+} iron is slow unless it is catalysed by the bacteria like *Acidithiobacillus ferrooxidans* at a favourable pH, <3.0 (Gleisner et al., 2006). The sulphate generated by the oxidation of pyrites acidifies the water and enters streams along with Witwatersrand, severely polluting the ground water and soils. The gold tailing impoundment in the Witwatersrand can typically be understood by the scheme presented in Figure 8.3. The study carried out by Naicker et al. (2003) revealed that the contamination of acidic ground water with the stream and soils varies seasonally, but interestingly, the surface water upstream from the mining area is found to be neutral and have low metal concentrations. Near to the mining area the pH of water falls to ~5 with a high concentration of sulphate and heavy metals therein. The analysis of the white colour surface crust developed along the river banks is gypsum with co-precipitated/adsorbed heavy metals onto it.

The polluting effect of acid mine drainage is particularly pronounced in the upper catchments of the Blesbokspruit and Klip Rivers, which draining the southern Witwatersrand escarpment. The water discharge of the accumulated volume in the voids of closed gold mines on the Witwatersrand to neighbouring mines generally of low quality is necessitating basic additional treatment by lime to raise the pH with air blowing to oxidize and

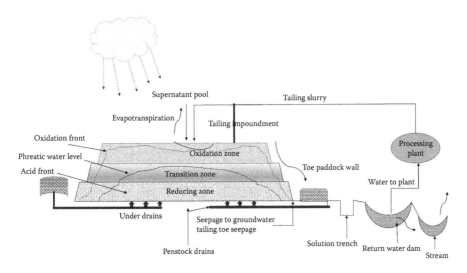

FIGURE 8.3
A schematic of a typical gold tailings impoundment in the Witwatersrand. (Modified and Adopted from Hansen, 2015)

precipitate the iron and other heavy metals. The iron is allowed to settle and is disposed of on tailings dumps while discharging the water into local rivers. Although the water discharge has a neutral pH, it also has a very high sulphate (~1500 mg/L) concentration and thus adds more pollution to the load already carried by the rivers in mining areas. The pollution arising from gold mines of the Central and Western basins is well illustrated by the salinity of the Vaal River; a periodic release of water from Vaal Dam significantly reduces the salinity for downstream Vaal River users.

In a fast-changing world, stringent environmental rules and global thirst for achieving sustainability in any field, the negligence of human perception cannot be affordable in the long run. It is time to recognize the need of people connected with a safe and sustainable operation of gold metallurgy. After such case studies, companies are trying to change their practices in a positive direction. By respecting human rights as a fundamental operating principle and recognizing the dignity of the miners/workers and other actors involved in gold mining and metallurgy operations, Barrick Gold has developed a compliance program (as represented in Figure 8.4) to help embed ethical b›aviour and respect for all employees and partners (http://www.barrick.com/responsibility/society/human-rights/default.aspx). Grounded in global standards in accordance with sustainable development; leading from the top, and embedded throughout the organization, shared learning, partnership, and collaboration are the key factors identified by the company. To mitigate

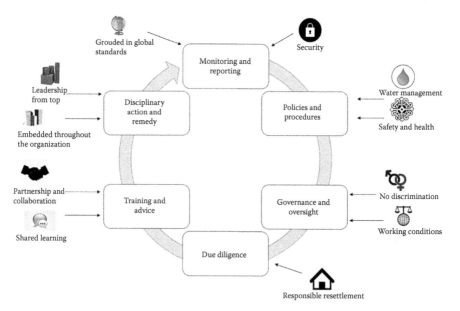

FIGURE 8.4

An example of the compliance program developed by Barrick Gold based on practising the human rights of workers. (Modified and Adopted from the online source material, http://www.barrick.com/responsibility/society/human-rights/default.aspx)

the negative impacts, the systematic elements of policies and procedures, governance and oversight, due diligence, training and advice, disciplinary action and remedy, and monitoring and reporting are impacted throughout the entire operation.

References

Akabzaa, T. 2000. *Boom and Dislocation: The Environmental and Social Impacts of Mining in the Wassa West District of Ghana*. Third World Network, Accra, Ghana.

Akabzaa, T. 2001. Research for advocacy on issues on mining and the environment in Africa: A case study of the Tarkwa mining district, Ghana. In: *Mining, Development and Social Conflicts in Africa*. Third World Network, Accra, Ghana.

Akabzaa, T., Darimani, A. 2001. The impact of mining sector investment: A study of the Tarkwa mining region. Draft report. Structural Adjustment Participatory Review Initiative (SAPRI), Accra, Ghana.

Akagi, H., Castillo, E.S., Cortes-Maramba, N., Francisco-Rivera, A.T., Timbang, T.D. 2000. Health assessment for mercury exposure among schoolchildren residing near a gold processing and refining plant in Apokon, Tagum, Davao del Norte, Philippines. *Science of the Total Environment*. 259(1): 31–43.

Akagi, H., Malm, O., Branches, F., Kinjo, Y., Kashima, Y., Guimaraes, J.R.D., Kato, H. 1995. Human exposure to mercury due to goldmining in the Tapajos river basin, Amazon, Brazil: Speciation of mercury in human hair, blood and urine. In: Porcella, D.B., Huckabee J.W., Wheatley B. (Eds), *Mercury as a Global Pollutant*. Springer, Dordrecht, The Netherlands.

Akpalu, W., Parks, P. 2007. Natural resource use conflicts: Gold mining in tropical rainforest in Ghana. *Environment and Development Economics* 12: 55–72.

Amenumey, D.E.K. 2008. *Ghana. A Concise History from Pre-Colonial Times to the 20th Century*. Woeli Publishing Services, Accra, Ghana.

Appleton, J. D., Williams, T. M., Breward, N., Apostol, A., Miguel, J., Miranda, C. 1999. Mercury contamination associated with artisanal gold mining on the island of Mindanao, the Philippines. *Science of the Total Environment*. 228(2): 95–109.

Aryeetey, E., Kanbur, R. 2008. *The Economy of Ghana. Analytical Perspectives on Stability, Growth and Poverty*. James Currey/Woeli Publishing Services, New York, Accra.

Ayine, D. 2001. The human rights dimension to corporate mining in Ghana: The case of Tarkwa district. In: *Mining, Development and Social Conflicts in Africa*. Third World Network, Accra, Ghana.

Barbosa, A.C., Silva, S.R.L., Dórea, J.G. 1998. Concentration of mercury in hair of indigenous mothers and infants from the Amazon basin. *Archives of Environmental Contamination and Toxicology*. 34(1): 100–105.

Burrow, E., Bird, L. 2017. Gold, guns and China: Ghana's fight to end galamsey. In: *African Arguments*. http://africanarguments.org/2017/05/30/gold-guns-and-china-ghanas-fight-to-end-galamsey/.

Chandra, A.P., Gerson, A.R. 2010. The mechanisms of pyrite oxidation and leaching: A fundamental perspective. *Surface Science Reports*. 65(9): 293–315.

CHRAJ (Commission on Human Rights and Administrative Justice). 2008. The state of human rights in mining communities in Ghana. The Commission on Human Rights and Administrative Justice, Accra, Ghana.

de Lacerda, L.D., Salomons, W. 1998. *Mercury from Gold and Silver Mining. A Chemical Time Bomb?* Springer Verlag, Berlin, Gemany.

Drasch, G., Böse-O'Reilly, S., Beinhoff, C., Roider, G., Maydl, S. 2001. The Mt. Diwata study on the Philippines 1999—Assessing mercury intoxication of the population by small scale gold mining. *Science of the Total Environment.* 267: 151–168.

Eisler, R. 2003. Health risks of gold miners: A synoptic review. *Environmental Geochemistry and Health.* 25: 325–345.

Garvin, T., McGee, T.K., Smoyer-Tomic, K.E., Aubynn, E.A. 2009. Community—Company relations in gold mining in Ghana. *Journal of Environmental Management.* 90: 571–586.

Ghana Chamber of Mines. 2007. *Publish What You Pay.* Accra, Ghana.

Ghanaian Chronicle. 1998. *Fear, Panic Grip Tarkwa and Environs.* May 28.

Gleisner, M., Herbert, R.B., Kockum, P.C.F. 2006. Pyrite oxidation by *Acidithiobacillus ferrooxidans* at various concentrations of dissolved oxygen. *Chemical Geology.* 225(1): 16–29.

Hansen R.N. 2015. Contaminant leaching from gold mining tailings dams in the Witwatersrand basin, South Africa: A new geochemical modelling approach. *Applied Geochemistry.* 61: 217–223.

Hilson, G. 2002. Harvesting mineral riches: 1000 years of gold mining in Ghana. *Resources Policy.* 28: 13–26. http://www.barrick.com/responsibility/society/human-rights/default.aspx

Krig, H.A., Malm, O., Akagi, H. 1997. Methylmercury in hair samples from different riverine groups, Amazon, Brazil. *Water, Air, and Soil Pollution.* 97(1–2): 17–29.

Lin, Y., Guo, M., Gan, W. 1997. Mercury pollution from small gold mines in China. *Water, Air, and Soil Pollution.* 97(3–4): 233–239.

Malm, O., Pfeiffer, W.C., Souza, C.M., Reuther, R. 1990. Mercury pollution due to gold mining in the Madeira River basin, Brazil. *Ambio.* 19(1): 11–15.

Marsden, D.D. 1986. The current limited impact of Witwatersrand gold-mine residues on water pollution in the Vaal River system. *Journal of the Southern African Institute of Mining and Metallurgy.* 86(12): 481–504.

Naicker, K., Cukrowska, E., McCarthy, T.S. 2003. Acid mine drainage arising from gold mining activity in Johannesburg, South Africa and environs. *Environmental Pollution.* 122(1):29–40.

Nyame, F., Grant, J.A., Yakovleva, N. 2009. Perspectives on migration patterns in Ghana's mining industry. *Resource Policy.* 34: 6–11.

Ogola, J.S., Mitullah, W.V., Omulo, M.A. 2002. Impact of gold mining on the environment and human health: A case study in the Migori gold belt, Kenya. *Environmental Geochemistry and Health.* 24(2): 141–157.

Papworth, S., Rao, M., Oo, M.M., Latt, K.T., Tizard, R., Pienkowski, T., Carrasco, L.R. 2017. The impact of gold mining and agricultural concessions on the tree cover and local communities in northern Myanmar. *Scientific Reports.* 7: 46594. DOI:10.1038/srep46594.

Sana, A., Brouwer, C.D., Hien, H. 2017. Knowledge and perceptions of health and environmental risks related to artisanal gold mining by the artisanal miners in Burkina Faso: A cross-sectional survey. *Pan African Medical Journal.* 27: 280. DOI:10.11604/pamj.2017.27.280.12080.

Santa Rosa, R.M., Müller, R.C., Alves, C.N., Sarkis, J.E.D.S., Bentes, M.H. D.S., Brabo, E., de Oliveira, E.S. 2000. Determination of total mercury in workers' urine in gold shops of Itaituba, Para State, Brazil. *Science of the Total Environment.* 261(1): 169–176.

World Gold Council Report. 2013. The gold mining industry: Reputation and issues. A survey of senior stakᐤolders and opinion formers. https://globescan.com/the-gold-mining-industry-reputation-and-issues-highlights-report/.

Index

Page numbers followed by *f* indicate figures; those followed by *t* indicate tables.

Milton Keynes UK
Ingram Content Group UK Ltd.
UKHW040102071024
449327UK00019B/741

9 780367 572075